158298

KV-577-601

STRATHCLYDE UNIVERSITY LIBRARY

30125 00090758 3

ANDERSONIAN LIBRARY
★
WITHDRAWN
FROM
LIBRARY
STOCK
✦
UNIVERSITY OF STRATHCLYDE

Books are to be returned on or before
the last date below

11. OCT 1983

24. 1982

14. JUL 1983

2 2 JAN 1992

1 1 MAR 1992

5 MAR 1998

−5. APR 1984

11. JUN 1985

5 JAN 1988

0 4 APR 1996

3 0 NOV 1992

2 9 MAR 1993

1 5 JUN 1994

LIBREX —

BUILDING SCIENCE AND MATERIALS

John Elliott

B.Sc., M.Phil., C.Chem, M.R.I.C.

UNIVERSITY OF
STRATHCLYDE LIBRARIES

© John Elliott 1977

All rights reserved. No part of this publication may be reproduced
or transmitted, in any form or by any means, without permission.

First published 1977 by
THE MACMILLAN PRESS LTD
London and Basingstoke
Associated companies in Delhi Dublin
Hong Kong Johannesburg Lagos Melbourne
New York Singapore and Tokyo

Typeset in 10/12 Times
and printed in Great Britain by A. Wheaton & Co, Ltd.

British Library Cataloguing in Publication Data

Elliott, John
  Building science and materials.–(Macmillan
technician series).
  1. Building
  I. Title   II. Series
  690   TH145

  ISBN 0–333–21489–7

This book is sold subject to the standard conditions of the
Net Book Agreement.

The paperback edition of this book is sold subject to the
condition that it shall not, by way of trade or otherwise, be
lent, resold, hired out, or otherwise circulated without the
publisher's prior consent in any form of binding or cover other
than that in which it is published and without a similar
condition including this condition being imposed on the
subsequent purchaser.

Contents

Foreword

This book is written for one of the many technician courses now being run at technical colleges in accordance with the requirements of the **Technician Education Council** (TEC). This Council was established in March 1973 as a result of the recommendation of the Government's Haslegrave Committee on Technical Courses and Examinations, which reported in 1969. TEC's functions were to rationalise existing technician courses, including the City and Guilds of London Institute (C.G.L.I.) Technician courses and the Ordinary and Higher National Certificate courses (O.N.C. and H.N.C.), and provide a system of technical education which satisfied the requirements of 'industry' and 'students' but which could be operated economically and efficiently.

Four qualifications are awarded by TEC, namely the Certificate, Higher Certificate, Diploma and Higher Diploma. The **Certificate** award is comparable with the O.N.C. or with the third year of the C.G.L.I. Technician course, whereas the **Higher Certificate** is comparable with the H.N.C. or the C.G.L.I. Part III Certificate. The **Diploma** is comparable with the O.N.D. in Engineering or Technology, the **Higher Technician Diploma** with the H.N.D. Students study on a part-time or block-release basis for the Certificate and Higher Certificate, whereas the Diploma courses are intended for full-time study. Evening study is possible but not recommended by TEC. The Certificate course consists of fifteen Units and is intended to be studied over a period of three years by students, mainly straight from school, who have three or more C.S.E. Grade III passes or equivalent in appropriate subjects such as mathematics, English and science. The Higher Certificate course consists of a further ten Units, for two years of part-time study, the total time allocation being 900 hours of study for the Certificate and 600 hours for the Higher Certificate. The Diploma requires about 2000 hours of study over two years, the Higher Diploma a further 1500 hours of study for a further two years.

Each student is entered on to a **Programme** of study on entry to the course; this programme leads to the award of a Technician Certificate, the title of which reflects the area of engineering or science chosen by the student, such as the Telecommunications Certificate or the Mechanical Engineering Certificate. TEC have created three main **Sectors** of responsibility

Sector A responsible for General, Electrical and Mechanical Engineering

Sector B responsible for Building, Mining and Construction Engineering

Sector C responsible for the Sciences, Agriculture, Catering, Graphics and textiles.

Each sector is divided into programme committees, which are responsible for the specialist subjects or programmes, such as A1 for General Engineering, A2 for Electronics and Telecommunications Engineering, A3 for Electrical Engineering, etc. Colleges have considerable control over the content of their intended programmes, since they can choose the Units for their programmes to suit the requirements of local industry, college resources or student needs. These Units can be written entirely by the college, thereafter called a college-devised Unit, or can be supplied as a Standard Unit by one of the programme committees of TEC. **Assessment** of every Unit is carried out by the college and a pass in one Unit depends on the attainment gained by the student in his coursework, laboratory work and an end-of-Unit test. TEC moderate college assessment plans and their validation; external assessment by TEC will be introduced at a later stage.

The three-year Certificate course consists of fifteen Units at three Levels: I, II and III, with five Units normally studied per year.

Entry to each Level I or Level II Unit will carry a prerequisite qualification such as C.S.E. Grade III for Level I or O-level for Level II; certain Craft qualifications will allow students to enter Level II direct, one or two Level I units being studied as 'trailing' Units in the first year. The study of five Units in one college year results in the allocation of about two hours per week per Unit, and since more subjects are often to be studied than for the comparable City and Guilds course, the treatment of many subjects is more general, with greater emphasis on an **understanding** of subject topics rather than their application. Every syllabus to every Unit is far more detailed than the comparable O.N.C. or C.G.L.I. syllabus, presentation in **Learning Objective** form being requested by TEC. For this reason a syllabus, such as that followed by this book, might at first sight seem very long, but analysis of the syllabus will show that 'in-depth' treatment is not necessary—objectives such as ' . . . **states** Ohm's law . . .' or ' . . . **lists** the different types of telephone receiver . . .' clearly do **not** require an understanding of the derivation of the Ohm's law equation or the operation of several telephone receivers.

This book satisfies the learning objectives for one of the many TEC Standard Units, as adopted by many technical colleges for inclusion into their Technician programmes. The treatment of each topic is carried to the depth suggested by TEC and in a similar way the **length** of the Unit (sixty hours of study for a full Unit), **prerequisite qualifications, credits for alternative qualifications** and **aims of the Unit** have been taken into account by the author.

Preface

This book was written primarily for students on the new TEC Construction courses, although it is also suitable for O.N.C./O.N.D., H.N.D. and Bridge courses in Construction, F.T.C. and Link courses.

It is felt that there is a need for a book such as this to supplement lectures on the subject, but its size has deliberately been kept compact in the interest of economy, since science is not generally required on courses above TC or O.N.D. in Construction. For this reason, experiments and some elementary material have been omitted.

I am very grateful to Mrs Sylvia Gould, for typing the manuscript.

JOHN ELLIOTT

1 Plastics

A revolution has occurred in the past hundred years, brought about by the introduction of plastics materials, which are synthetic and man-made. Other materials, such as metal, wood and concrete, have been known for centuries, but plastics materials were widely produced for the first time only relatively recently, and are the fruits of modern science and technology.

Examples

Polystyrene   Polyesters
Nylon   Polytetrafluoroethylene
                                   (PTFE)

Polyethylene   Polyvinyl acetate (PVA)
Polypropylene   Polymethyl methacrylate
Polyvinyl chloride (PVC)   Phenol formaldehyde
Cellulose acetate   Urea formaldehyde
Casein plastic   Melamine formaldehyde

How Plastics are Made

Quite a number of ingredients are needed to make a fruit cake, and in the same way a plastics material is made up of a mixture of different things. A black plastics macintosh contains the essential *polymer* (in this case usually PVC), a black dye and a plasticiser (something added to make the material flexible). A floor tile may contain some sand to make it more hard-wearing; a heat-resistant article may contain asbestos, and so on.

Polymers

When atoms join up, *molecules* are formed. Molecules can be either large or small.

A molecule containing hundreds or thousands of atoms is called a *giant molecule* or *polymer*.

Polyethylene or *polythene* (figure 1.1) is composed of giant molecules, each molecule being a long chain made up of carbon and hydrogen atoms.

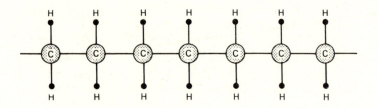

Figure 1.1

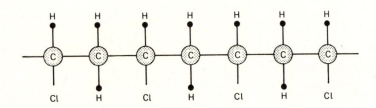

Figure 1.2

Polyvinyl chloride or *PVC* (figure 1.2) is also composed of giant long-chain molecules, containing atoms of carbon, hydrogen and chlorine.

Every plastics material contains giant molecules, although not necessarily in chain form, because sometimes the chains join up by linking across to each other, forming a three-dimensional network.

'Thermoplastics' and 'Thermosets'

A plastics bucket may be formed by a moulding process. The material is heated until it is molten, and is then forced into a cavity in the shape of a bucket. On cooling, the plastics material solidifies. The bucket can later be melted down again and remoulded into something else. This is not possible with wood or concrete. The material is *thermoplastic*, that is, it can be repeatedly melted down and remoulded. However, with some plastics materials, this does not happen. After the first moulding, it is no longer possible to melt it down and remould it again. On reheating, the material simply chars or burns. These materials are *thermosets*.

Generally, thermoplastics contain long-chain polymers, which can writhe and slide across each other on heating (like spaghetti), whereas thermosets contain a much more rigid three-dimensional cross-linked network.

Thermoplastics	*Thermosets*
Polystyrene	Phenol formaldehyde
Nylon	Urea formaldehyde
Polyethylene	Melamine formaldehyde
PVC	Some polyesters
Polypropylene	
Polyesters	

Theory of how Polymers are Made: Polymerisation

Chemists have discovered that the best way of making polymers (or giant molecules) is by joining smaller molecules together. Not all small molecules are suitable, for instance, water molecules (H_2O) cannot be used to make polymers. Substances which do contain suitable molecules are called *monomers*. A gas called ethylene (figure 1.3) contains small molecules which can be joined up to make large molecules.

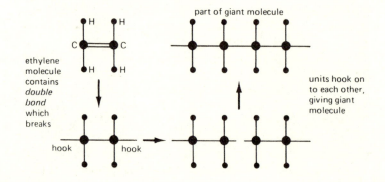

Figure 1.3

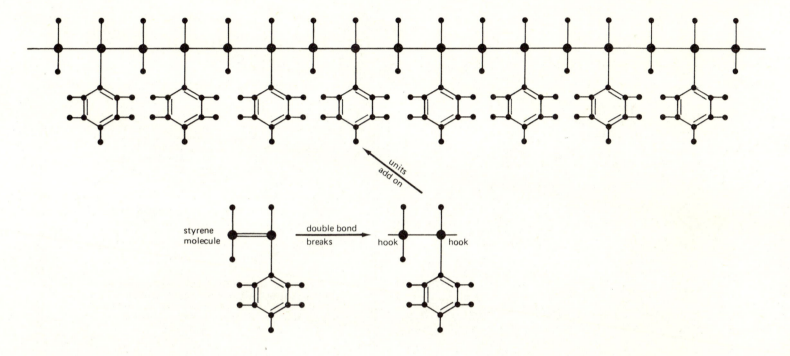

Figure 1.4

The above process is called *addition polymerisation*, since the polymer is made by adding together molecules. Polymers made this way include polythene, PVC, polymethyl methacrylate, and polystyrene, shown in figure 1.4.

A more complicated but efficient process for making polymers is by a process called *condensation polymerisation*, which again operates by joining up small molecules. The process is illustrated by the example in figure 1.5.

As long as there is a hooking unit (OH or COOH) at each end of the chain, the molecule can be made as large as desired. In this process (condensation polymerisation) a polymer is always produced together with small molecules, such as water.

Practical Polymer-making (figure 1.6)

Compounding

Polystyrene is a polymer, made by the above process. If the material is to be made into a bread bin, for example, various additives are incorporated, such as dyes, fillers, plasticisers and stabilising agents. The end result is a plastics materials. The process of mixing the ingredients together to make them homogeneous is called *compounding* (figure 1.7).

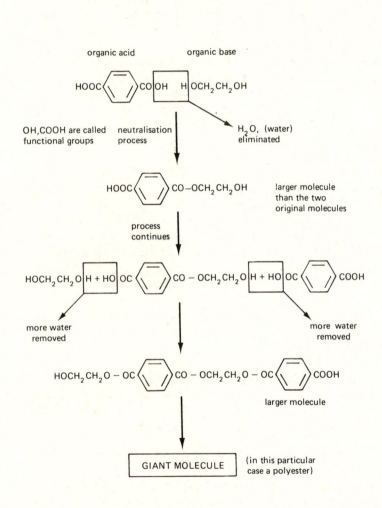

Figure 1.5

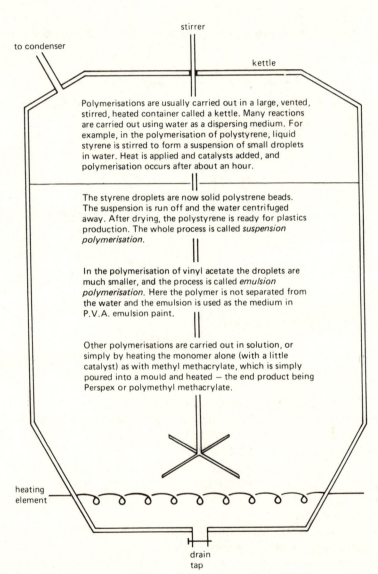

Figure 1.6

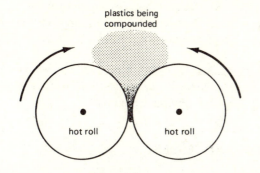

plastics being compounded

hot roll   hot roll

Figure 1.7

Summary

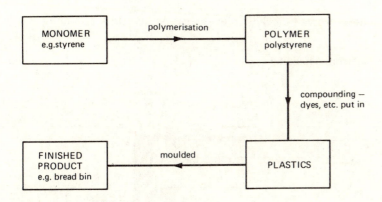

| MONOMER e.g.styrene | → polymerisation → | POLYMER polystyrene |

compounding — dyes, etc. put in

| FINISHED PRODUCT e.g. bread bin | ← moulded ← | PLASTICS |

Figure 1.8

PROPERTIES OF PLASTICS

Thermoplastics

Thermoplastic materials are polymers or giant molecules com-

posed of long chains. In other words they are linear polymers. As such they have been adapted to produce strong fibres, for example, nylon, polyester, polypropylene, acrylics and others, have all been spun into fibres and woven into cloth, or made into ropes, and so on.

An outstanding feature about thermoplastics is their ability to melt down and remould into some other shape. Plasticisers, which are usually inert oily liquids, can be added to make the material more flexible. Whether plasticiser has been added or not, the surfaces of thermoplastics tend to be soft and easily scratched.

Most thermoplastics have densities of about the same value as water, unless they contain heavy fillers, such as sand. An exception to this is polytetrafluoroethylene (PTFE), which has a density approaching that of concrete.

In general, thermoplastics have poor fire resistance and are inflammable, unless asbestos powder has been used as a filler.

As regards weathering, thermoplastics are durable, water-resistant and water-insoluble, but some are prone to discolouring on prolonged exposure to sunlight. PVC in particular yellows unless a light stabiliser such as dibutyl tin dilaurate is incorporated.

Most thermoplastic polymers are white or colourless, and are therefore capable of taking dyes, to produce a whole range of attractive products.

Mechanical properties vary from one material to another; for instance, polystyrene has poor impact strength, whereas that of polythene and polypropylene is good. However, all thermoplastics suffer badly from creep characteristics; that is, they deform permanently under stress, and have little resistance to loading.

Thermosets

Thermosetting materials are composed of giant three-dimensional cross-linked molecules. This structure has harder and more rigid properties than thermoplastics. Thermoset surfaces tend to be harder, for example, Formica tops are more scratch-resistant. Plasticisers are not normally added.

The fire resistance of thermosets, like thermoplastics, tends to be poor, but its behaviour is rather like that of wood; that is, it chars and burns, but does not become molten.

Thermoset materials are, on the whole, mechanically stronger than thermoplastics and are not subject to creep, but they are more brittle.

As regards colour, the same range is not so generally available, because the raw materials for making thermosets are sometimes coloured, and darkening also tends to occur in the polymerisation process.

Some thermosets, particularly phenolics, are tremendously important as agents for bonding together paper and wooden laminates to produce boards and other articles, which have considerable mechanical strength, and they find many applications in the construction industry. The electrical insulation properties of these laminates is very good, and many applications are found in the electrical industry.

ADDITIVES USED WITH PLASTICS MATERIALS

As we have already stated, various additives are mixed in or compounded with the appropriate polymer to give a plastics with the required properties. The following is a list of some of the additives.

(1) *Fillers*, such as sand, sawdust, chalk, asbestos powder. These impart certain properties to the article being made: asbestos for fire resistance, sand for abrasion resistance, and so on.

(2) *Extenders*, such as cork dust. These materials are added purely to increase bulk and reduce cost. They do not usually modify the properties of the plastics material involved.

(3) *Pigments*, such as iron oxide ⎫
(4) *Dyes*   ⎬ impart colour

(5) *Plasticisers*, such as dibutyl phthalate. These are inert, tasteless, non-toxic, involatile organic liquids, which are compatible with plastics, and increase the flexibility of the article being made. Thus plastics down-pipes are made from rigid PVC, containing no plasticiser, whereas PVC tablecloths contain a fairly high proportion of plasticiser.

(6) *Cross-linking agents*. Certain linear polymers can be converted into three-dimensional cross-linked polymers by linking across the chains. Cross-linking agents are substances which perform this task; an example is styrene, used to cross-link polyesters.

(7) *Antioxidants*   ⎫ Substances introduced to prevent
(8) *Light stabilisers* ⎬ decomposition by oxygen and light
(9) *Perfume*, for obvious reasons!

MOULDING PROCESSES

Many processes are available for moulding plastics materials. The most important ones are as follows.

> *Injection moulding*
> *Thermo- and vacuum forming*
> *Extrusion*
> *Compression moulding*
> *Calendering*

Other processes include blow extrusion and slush moulding.

Injection Moulding

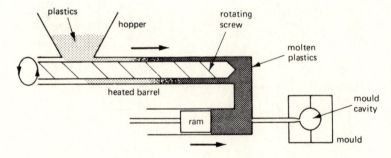

Figure 1.9

Operation (see figure 1.9)

(1) Plastics chips put into hopper.

(2) Material moves forward by screw action and also melts.

(3) Material moves into ram compartment.

(4) Ram moves forwards and injects material into mould cavity.

(5) On cooling, the mould is opened and the plastics moulding is removed.

(6) Process is repeated.

(In some machines the screw also serves as the ram.)

Applications

A very wide range of articles is produced, such as combs, lavatory seats, door-handles and drain-pipe brackets.

Thermo- and Vacuum Forming

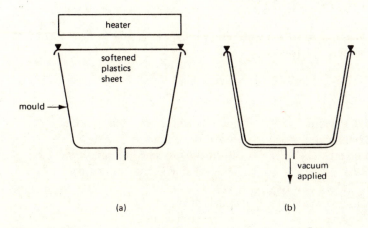

Figure 1.10

Operation (see figure 1.10)

(1) Figure 1.10a: plastics sheet clamped in position and softened by heater.

(2) Figure 1.10b: heater removed; plastics sheet drawn down into mould by vacuum.

(3) On cooling, the moulding is removed.

Applications

The advantage of the process is the simplicity and cheapness of the equipment. Used for making disposable plastics cups, etc.

Extrusion

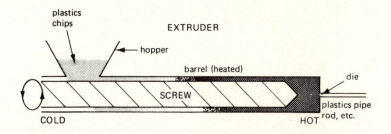

Figure 1.11

Operation (see figure 1.11)

The rotating screw pushes the plastics material forward. The material melts towards the front of the extruder and is forced through a shaped hole called a *die*. Cooling is effected by cold air or water.

Applications

Any article with a continuous section can be produced by this method, for example, hosepipe, drain-pipe, guttering, curtain rail and diffusers for tubular lighting. Plastics-covered wire is made this way, as are plastics bottles (made by extruding tube and then blowing it while still hot into a bottle mould).

Compression Moulding

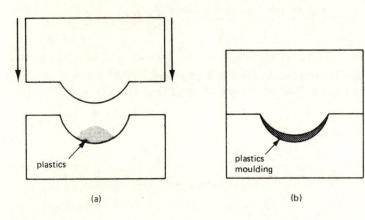

(a)   *(b)*

Figure 1.12

Operation (see figure 1.12)

(1) Figure 1.12a: mould closes.

(2) Figure 1.12b: application of heat and pressure causes plastics to melt and flow into mould cavity.

(3) Mould opens and moulding is removed.

Applications

Shallow articles, for example, plates and saucers; particularly suitable for thermosets.

Calendering

Operation (see figure 1.13)

Plastics material is rolled out into sheets up to 3 m wide.

Applications

Plastics sheet, particularly PVC, for tablecloths, macintoshes, etc.

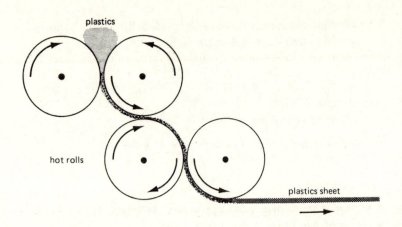

Figure 1.13

TESTING PLASTICS

A useful test for identifying plastics is simply to *heat* them and observe the following.

Material	On Heating
Polystyrene	Sweet smell, burns, black smoke
Nylon	Smell of celery
Polyethylene	Burns, drips, smell of candlewax
PVC	Acidic fumes—turn litmus red
Cellulose acetate	Smell of vinegar
Casein plastic	Smell of burnt cheese
Perspex	Sweet fruity smell
Bakelite	Chars and burns, sometimes fishy smell

EXERCISES

1.1 Suggest a suitable plastics material for each of the following articles and indicate by which process it may be produced.

Articles	Plastics Used	How Made
Guttering clip	PVC	Injection moulding
Cavity fill		
Curtain rail		
Electric switch		
Damp-proof sheet		
Bath		
Roof-Light		
Lampshade		
Floor-tile		
Mains water pipe		
Toilet seat		
Acoustic ceiling tile		
Guttering		
Wall tile		
Door-hinge		
Electric wiring		
Sarking felt		
Coldwater cistern		
Door furniture		
'Rubber' gloves		

1.2 List six general properties and characteristics of thermo-softening plastics, and six of thermosetting plastics materials. Outline the method of manufacture and uses of *two* types of plastics materials used in building.

1.3 A company manufacturing plastics constructional articles is asked to supply the following items
    (a) thin sheeting to be put under floor screeding for damp-proofing

  (b) guttering and down-pipes
  (c) baths.
Select and name a suitable plastics material for each and draw simple diagrams to show how the items may be made.

1.4 (a) Explain the meaning of the following terms when used in plastics technology, and give an example in each case.
    (i) cross-linked polymer
    (ii) monomer
    (iii) plasticiser
    (iv) addition polymerisation.

  (b) Suggest *one* type of plastics material for each of the following articles, and describe how the articles may be made:
    (i) door-hinges
    (ii) roof-light domes
    (iii) foamed ceiling-tiles.

1.5 Give an account of the nature of thermosetting and thermo-softening plastics used in building, listing the general characteristics of each type and including the principal molecular difference. Also outline the manufacture, properties and uses of polystyrene.

1.6 Give a general account of plastics materials, indicating those in general use, and the methods of processing employed. Emphasise their applications to the building industry.

2 Metals and Corrosion

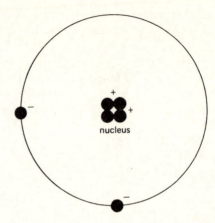

Figure 2.1

All matter is made of atoms (figure 2.1). All atoms are composed of *protons, neutrons* and *electrons*.

proton has positive charge and mass 1 unit
neutron has no charge and mass of 1 unit
electron has negative charge and mass which is negligible

The *nucleus* of the atom is composed of protons and neutrons. Electrons revolve around the nucleus in orbits or shells. (This model of the atom, which was conceived by Niels Bohr, is now somewhat outdated, and the atom is now considered to be more complex. However, the Bohr atom is still used, since it gives a clear explanation of many phenomena.)

In an atom there is always an equal number of protons and electrons—the number of neutrons varies. For example, in figure 2.1, which shows an atom of helium, there are 2 protons and 2 electrons; there are also in this case, by coincidence, 2 neutrons. In an atom of iron, there are 26 protons and 26 electrons. In each case an equal number of positive and negative charges maintains a neutral atom.

Sometimes an atom loses or gains electrons, giving an imbalance of numbers of positive protons and negative electrons. Thus, if an iron atom Fe loses 2 electrons, there is a residual double-positive charge on the atom, which is now written Fe^{2+}. This charged atom is called an *ion*.

$$Fe \rightarrow Fe^{2+} + 2 \text{ electrons}^{2-}$$
iron atom   *iron ion*

Some atoms gain extra electrons and become negative ions. Thus a chlorine Cl atom can pick up an electron, giving rise to a chlorine ion Cl^-.

$$Cl + 1 \text{ electron}^- \rightarrow Cl^-$$
chlorine atom   *chlorine ion*

All metal atoms give positive ions. Positive ions are called *cations* and negative ions *anions*.

When metal atoms are in an *ionic* condition, they are very reactive and subject to decomposition by chemical change. It is considered that all corrosion by metals takes place in this manner.

Metal	Ions Produced
Iron (Fe)	Fe^{2+} (ferrous): Fe^{3+} (ferric)
Copper (Cu)	Cu^+ (cuprous): Cu^{2+} (cupric)
Lead (Pb)	Pb^{2+} (plumbous): Pb^{4+} (plumbic)
Zinc (Zn)	Zn^{2+}
Aluminium (Al)	Al^{3+}

METALS USED IN CONSTRUCTION

Iron, steel, copper, lead, zinc and aluminium are all used extensively in the construction industry and, unfortunately, some are very prone to chemical attack from the atmosphere, unlike timber, glass, ceramics and plastics. The metals are gradually eroded away into their respective compounds, which do not have as good constructional properties as the parent metal. One only has to look under an old car to see what an excellent crop of iron oxide is available from half a ton of metal.

SPEED OF CORROSION

The speed at which metals corrode depends on how quickly they can form ions and the *Activity Series*, or *Electrochemical Series*, is a table of metals arranged in this fashion.

These metal are destroyed by water in a few minutes	Potasium, K Calcium, Ca Sodium, Na	These metals ionise most easily and therefore corrode most easily
Metals above here corrode more quickly than iron	Magnesium, Mg Aluminium, Al Zinc, Zn Iron, Fe Nickel, Ni Tin, Sn Lead, Pb (Hydrogen, H) Copper, Cu	Decrease in corrosion rate ↓
Very little corrosion	Mercury, Hg Silver, Ag	Ionise least easily therefore corrode least quickly
Used in jewellery	Gold, Au	

Cell Action

Notice what happens when two dissimilar metals (say copper and zinc) are placed in contact in a solution of electrolyte (liquid which conducts electricity)—see figure 2.2.

The conclusion we draw from these phenomena is that if two dissimilar metals are connected in an electrolyte, the corrosion rate of the more active metal is increased, while the corrosion rate of the less active metal is decreased.

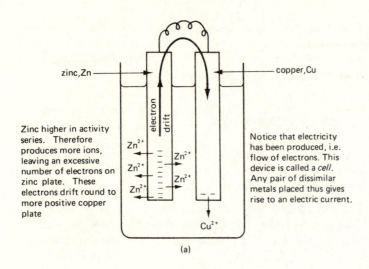

Zinc higher in activity series. Therefore produces more ions, leaving an excessive number of electrons on zinc plate. These electrons drift round to more positive copper plate

Notice that electricity has been produced, i.e. flow of electrons. This device is called a *cell*. Any pair of dissimilar metals placed thus gives rise to an electric current.

(a)

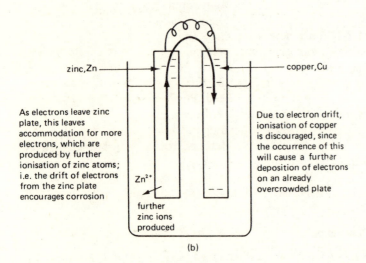

As electrons leave zinc plate, this leaves accommodation for more electrons, which are produced by further ionisation of zinc atoms; i.e. the drift of electrons from the zinc plate encourages corrosion

Due to electron drift, ionisation of copper is discouraged, since the occurrence of this will cause a further deposition of electrons on an already overcrowded plate

(b)

Figure 2.2

This can be used to advantage in the sacrificial method of preventing corrosion (see p. 13), and has a disadvantage in that two dissimilar metals connected together (such as copper and lead pipes in plumbing) tend to cause an increase in the corrosion of one of the metals.

Corrosion of Iron

Although the corrosion of iron is a very common phenomenon, its mechanism is very complex. The over-all process is shown in figure 2.3.

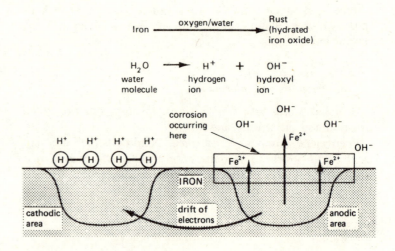

Figure 2.3

It is well known that both oxygen and water are necessary for iron to rust. The mechanism of rusting is thought to be as follows. At certain areas on the surface of iron in contact with air and moisture, iron atoms ionise and dissolve in the water (that is, they corrode). The electrons left on the surface drift away and congregate elsewhere on the surface. This process gives rise to what are known as *anodic* and *cathodic* areas.

Positive hydrogen ions (H^+) are attracted to the electrons in the cathodic area. Each ion picks up an electron and becomes an atom

$$H^+ + electron^- \rightarrow H \text{ atom}$$

Hydrogen atoms are very active and they quickly combine to form hydrogen molecules, which aggregate, giving bubbles of hydrogen gas

$$H + H \rightarrow H_2$$
$$\textit{atoms} \quad \textit{molecule}$$

This collection of bubbles around a localised spot is called *polarisation*.

Meanwhile at the anodic areas, hydroxyl ions (OH^-) are drawn towards the positive metal ions, and the following reaction occurs

$$Fe^{2+} + 2OH^- \rightarrow Fe(OH)_2$$
$$\textit{ferrous hydroxide}$$

In time the ferrous hydroxide is oxidised by oxygen to give ferric hydroxide, or hydrated ferric oxide, which is rust.

$$Fe(OH)_2 \xrightarrow{\text{oxygen}} Fe(OH)_3 \longrightarrow Fe_2O_3$$
$$\textit{rust}(\text{orange}) \quad \textit{old rust}(\text{brown})$$

The rate at which corrosion occurs depends on how quickly electrons are removed from the cathodic areas, because vacancies encourage further ionisation of iron atoms. Acidic water generally increases corrosion rates because of the large numbers of hydrogen ions present, which mop up large quantities of electrons.

Hydrogen bubbles disappear gradually, by combination with oxygen to form water

$$2H_2 + O_2 \rightarrow 2H_2O$$

PREVENTION OF CORROSION

The following is a summary of the methods employed for preventing corrosion of metals, particularly iron.

Stifling Action

The metal is coated with paint, grease, bitumen, etc., insulating the surface from the corroding atmosphere.

Use of Inhibitors

Inhibitors are generally oxidising agents which produce a protective coating on the metal surface. An example is phosphoric acid. When iron is dipped into the acid a chemical reaction occurs and iron phosphate is formed, which produces a protective, coherent, insulating layer.

Use of Alternative Materials

Materials such as plastics are not subject to corrosion.

Metals' Own Natural Protection

Some metals react chemically with the atmosphere and thus form a natural protective barrier on their surfaces.

(1) Aluminium forms a thin coherent layer of aluminium oxide on its surface.

(2) Copper forms a protective green coating of basic copper sulphate, for example roofs of the Kremlin and the Houses of Parliament, Ottawa.

(3) Stainless steel (an alloy of iron, chromium and nickel) forms an extremely thin but effective layer of iron oxide/chromium oxide.

(4) Lead gives a coherent film of lead oxide/carbonate.

(5) Zinc gives a coherent film of oxide, hydroxide and carbonate.

On the other hand, iron gives a crumbly non-protective, non-coherent film of iron oxide.

Sacrificial Action

When two dissimilar metals are in contact in an electrolyte, the metal higher in the activity series corrodes, or sacrifices itself, in preference to the other metal. Thus, if iron is to be protected, a metal higher in the series must be connected to it.

Many iron objects are galvanised, or coated with zinc, for this purpose—the zinc corrodes, or sacrifies itself, in preference to the iron. Another name for this type of corrosion prevention is *cathodic protection*. This method is often used in the construction industry to protect underground steel pipes (figure 2.4).

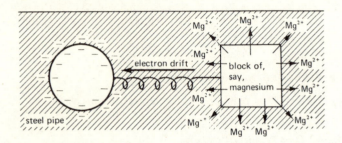

Figure 2.4

Magnesium ionises in preference to iron. The electrons left on the magnesium block by the ionisation drift across on to the steel pipe, reducing ionisation or corrosion. A modification of this method uses scrap iron instead of magnesium. To encourage the scrap iron to corrode in preference to the pipe (because they are, after all, both iron) an electric current is actually sent along the connecting wire to stimulate the electron drift. (figure 2.5).

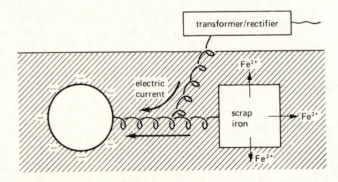

Figure 2.5

EXERCISES

2.1 Explain how iron rusts in the presence of water and oxygen, and give *four* methods used for preventing corrosion.

2.2 What is the explanation, in electrolytic terms, of corrosion? In what sense may aluminium be said to corrode?

Distinguish between the principles of the protection afforded to iron and steel in

(a) painting
(b) galvanising
(c) tin plating
(d) treatment with phosphoric acid.

2.3 Give an account of corrosion and its prevention in metals in

(a) buried iron and steel pipelines
(b) dissimilar metals in contact
(c) atmospheric corrosion of non-ferrous metals.

In each case, give two examples and explain why corrosion occurs.

2.4 What is meant by

(a) an ion
(b) the activity series for metallic elements?

Explain as fully as you can the rusting of iron by electrolytic processes. Give methods of prevention of rusting in iron.

Write down *six* uses of iron (or steel) in the construction industry.

2.5 (a) The protective coating to a buried steel pipeline is found to have small holes in it over a 150 m length of the pipe. With the aid of sketches show and explain *two* methods which can be used to protect the pipe and prevent the need for extensive repair or replacement of the coating. Give the advantages of the method generally.

(b) List *three* factors which could cause or accelerate the corrosion of the steel.

2.6 Outline, with the aid of sketches, *two* similar methods of preventing the corrosion of part of a steel structure immersed in water, where surface coating is impractical. Also with the aid of sketches describe experiments to prove the effectiveness of the above corrosion-protection methods.

3 Concrete

Concrete is obviously a very important constructional material. It is made by mixing together cement, aggregate (sand and/or gravel) and water, and its strength is tremendous. A whole new technology has been built up around concrete during the last hundred years or so, and its strength and suitability can now be tailored for a particular job to a fine degree. It is an easy matter to 'knock up' some concrete for a garden path, and a little extra water or sand makes no difference, but a concrete foundation for a multistorey building must be designed down to the last detail, and the civil engineers and technologists of today are very familiar with the intricacies of such a problem.

Concrete is an almost indispensable material. It is cheap, very strong and can last for tens or even hundreds of years in a suitable environment.

AGGREGATES

To be used in concrete, an aggregate must first of all be cheap and available in large quantities. It should be as strong mechanically as the resulting concrete in which it is used. Aggregates should not contain excessive dust and should be free from organic material such as dead leaves, which causes a reduction in strength and may produce unsightly stains.

If the concrete is to serve its most usual purpose, say, as a foundation of a building, sand and gravel aggregate are eminently suitable. For the production of lightweight concrete (see p. 25), lightweight aggregates are used. They are porous and less strong and are usually specially prepared. For heavyweight concrete, aggregates containing heavy metal atoms such as iron and barium are used. Aggregates can thus be classified broadly into three groups—normal, lightweight and heavy.

Normal Aggregares

(1) Natural aggregate: sand, crushed stone sand, crushed gravel sand, crushed gravel, crushed stone.

(2) Broken brick: generally from crushed engineering-grade brick.

(3) Blastfurnace slag: waste product obtained in the manufacture of iron.

Lightweight Aggregates

(1) Sintered pulverised fuel ash: made by mixing ash from power stations with water and heating to 1000 °C.

(2) Clinker: ash left from burning coal or coke.

(3) Foamed slag   ⎫
(4) Expanded clay ⎪ All made by heat treatment
(5) Expanded perlite ⎬ on the original raw material,
(6) Exfoliated vermiculite ⎪ for example, certain clays
(7) Expanded shale ⎭ on heating give off gases
                      which cause expansion.

(8) Pumice: material of volcanic origin (solidified lava).

Heavy Aggregates

(1) Barytes   ⎫
(2) Magnetite ⎪ All these materials are ores of
(3) Haematite ⎬ heavy metals, and are used for
(4) Limonite ⎪ making concrete for special
(5) Iron and Steel ⎭ purposes, for example, for
                   screening atomic piles.

MANUFACTURE OF CEMENT (see figure 3.1)

COMPOSITION OF CEMENT

All cements are composed of the oxides of metals, the chief ones being calcium oxide, CaO, silicon dioxide, SiO_2, aluminium oxide, Al_2O_3 and ferric oxide, Fe_2O_3. Note, however, that cement is not just a mixture of oxides; the molecules of these compounds combine in certain ways. The most common combination, for instance, is that of three CaO and one SiO_2 molecules connected together; the molecule is written $3CaO.SiO_2$. This substance is usually called *tricalcium silicate*, and almost half of ordinary Portland cement is made of it.

A combination of two molecules of CaO with one molecule of

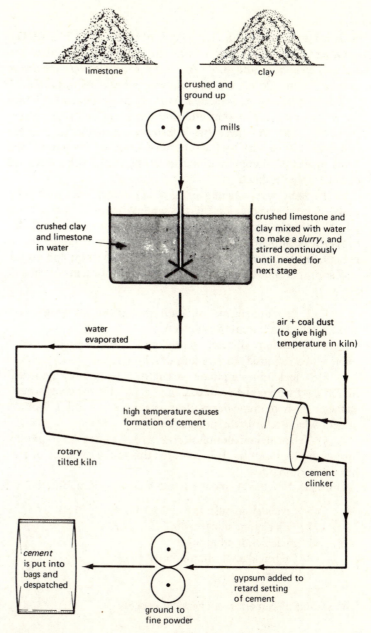

Figure 3.1

SiO_2 is also common, and is called *dicalcium silicate*, $2CaO.SiO_2$.

Other important combinations are $3CaO.Al_2O_3$, called *tricalcium aluminate*, and $4CaO.Al_2O_3.Fe_2O_3$, called *teretracalcium aluminoferrite*. Others are also known, but are less important.

You may be aware that sodium is a very dangerous metal, which sometimes catches fire when added to water, and chlorine is a very poisonous gas; yet these two substances combine to produce common salt—a very useful substance, essential for human life. Thus by chemical combination, the properties of substances are considerably modified.

In the same way, calcium oxide, CaO, (quicklime) is a white powder and silicon dioxide, SiO_2, is sand. When molecules of these join together, to give $3CaO.SiO_2$, the substance produced has properties completely different from the parent substances, and it is these properties which make concrete so important and useful.

The properties for these four important constituents of cement are as follows.

(1) Tricalcium silicate: on adding water and allowing to set, great mechanical strength is achieved.

(2) Dicalcium silicate: somewhat reduced strength, and a longer time required for strength development.

(3) Tricalcium aluminate: strength of the set or 'hydrated' material is further much reduced, but the setting time is very short. In fact a 'flash' or immediate setting occurs unless a little gypsum is added, which retards the process.

(4) Tetracalcium aluminoferrite: produces a material of only low strength on setting. It is coloured due to the presence of red-brown ferric oxide.

Ordinary Portland cement has the following composition

45% tricalcium silicate
27% dicalcium silicate
11% tricalcium aluminate
 8% tetracalcium aluminoferrite
 9% others

The setting of cement is a chemical process

cement + water (+ aggregate) → concrete + heat

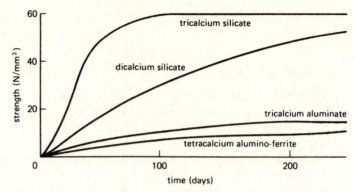

Strength Developments of Concrete Constituents

Figure 3.2

It is an exothermic reaction, that is, it gives out heat. The amount of heat evolved depends on the readiness for reaction or spontaneity of the reaction, and the length of time involved—in the case of concrete, about a year. It will be appreciated that tricalcium silicate evolves the most heat. Tricalcium aluminate also gives out much heat, but over a very short time, due to its flash setting.

TYPES OF CEMENT

Portland Cement (figure 3.3)

High Alumina Cement (Ciment Fondu)

Whereas Portland cement is made by heating together limestone and clay (silicon dioxide), high alumina cement is prepared by heating together limestone and bauxite (aluminium oxide) to a temperature of about 1600°C.

Its advantages over Portland cement are that although it has the same setting time, its final compressive strength is much greater, it is also refractory, having a high heat resistance, and it develops its strength quickly. It is more expensive than Portland cement. Large

Portland Cement

ordinary Portland cement (O.P.C.) good general-purpose cement	rapid-hardening Portland cement (R.H.P.C.) hardens more rapidly than O.P.C., more finely ground than O.P.C., slightly more tricalcium silicate present, used in cold weather or when time is important	extra rapid-hardening Portland cement this is R.H.P.C. containing an accelerator such as calcium chloride
	low-heat Portland cement contains less tricalcium silicate than O.P.C., hardens and evolves less heat during cure, used in mass concrete and where trapped heat would damage concrete	sulphate resisting cement contains less tricalcium silicate than O.P.C., therefore it is more sulphate resisting
white and coloured Portland cement white cement is made using china clay instead of ordinary clay, pigments used in coloured cement	masonry cement contains O.P.C. + plasticiser + inert filler, for brick-laying, etc.	Portland blast furnace cement made by heating P.C. clinker with blast furnace slag, good resistance to chemicals

Figure 3.3

amounts of heat are evolved during setting so lean mixes are advisable, and because it has been found to have unstable characteristics in the long term, it is not now favoured for structural work.

CONCRETE MIX DESIGN

It is possible nowadays to produce a concrete with a specified

strength. The strength referred to here is the load required to crush the concrete. The technology of designing a concrete for a specified purpose is called *concrete mix design*. All concrete mixes are designed basically to possess *two* properties: *maximum strength* and *maximum durability*.

When designing a concrete one must bear in mind the following points.

(1) *Materials must be suitable*: *water* should be ideally of a *drinkable* standard; *aggregates* should be clean and suitable; *cement* should be the correct type, for example, low heat Portland cement if large masses of concrete are involved.

(2) *Aggregate should be well graded (figure 3.4)*: there should be a good particle size distribution from coarse to fine aggregate, to reduce porosity to a minimum. The strength of the final concrete depends on how porous it is; if it is full of holes, like Rice Crispies, it will have a very poor strength! Dense concrete, that is, concrete with very few voids, has great strength.

coarse aggregate
(large pores)

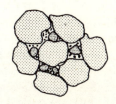

graded aggregate of large, medium and small particles (pores reduced to minimum)

Figure 3.4

(3) *Materials should be batched by weight:* batching is the process of measuring out the constituents of the concrete in their correct proportions before mixing. The only really accurate way of doing this is by weighing.

Batching by volume can be inaccurate for two reasons. Firstly aggregates tend to absorb moisture, the amount of which is hard to assess, and secondly, when sand absorbs water its volume increases by up to 25 per cent of its initial volume. This phenomenon is known as the *bulking of sand*, (figure 3.5) and is due to a film of water forming around the particle, increasing its effective volume. Thus the actual amount of sand in a given volume of damp sand is considerably less than the amount in its dry counterpart.

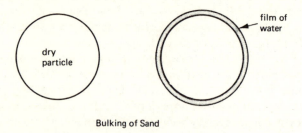

Bulking of Sand

Figure 3.5

(4) *Keep water/cement ratio as low as possible*: It is found that for the strongest concrete the water and cement should be in the proportion of about 1 to 3 by weight. However, it is soon found by practice that this mix, combined with graded aggregate, gives such a dry mixture that it is impossible to get much *compaction*. Its *workability* is said to be poor. Also, poor compaction means many voids or pores and this leads to a concrete of poor strength.

If more water is added, the mix becomes more workable and compaction is much better. In fact the more water that is added the better is the workability; but unfortunately only a certain amount of water is necessary for the chemical reaction or *hydration* with the cement, and the excess gradually evaporates, leaving voids which cause a strength reduction. It has been found that 1 per cent of interconnected void space produces a 5 per cent reduction in strength.

Thus a compromise has to be reached. The aim is to use the minimum amount of water to get the best workability and also maximum strength. It is worth mentioning here that a large excess

of fine aggregate should be avoided since this requires large amounts of water to make it workable with consequential decrease in concrete strength.

It can be shown by running a few tests that the water/cement ratio and compressive strength vary in the manner shown by the curve in figure 3.6. This relationship has been expressed by the Abram's equation, for air voids of less than 1 per cent

$$S = AB^{-x}$$

where x = water/cement ratio, S = strength of concrete and A and B are constants.

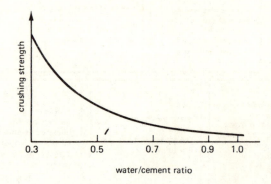

Figure 3.6

(5) *Aggregate/cement ratio is important*: The essential points here are that there must be sufficient cement to bind the aggregate together and the ratio must be such that the mix is workable.

(6) *Mix the ingredients well so that the mixture is homogeneous.*

(7) *Allow suitable conditions for curing:* This means allowing the concrete to dry out slowly at a reasonable temperature (10–30 °C), pouring on water if necessary in the initial stages, and/or covering with sacking.

CONCRETE TESTING (BS 1881)

The principal methods of testing concrete are

(1) Slump test
(2) Cube test
(3) Compaction factor test
(4) V–B consistometer test

Slump Test

The purpose of this test is to obtain some measure of the *workability* of the concrete being prepared. The test may be performed on site by a careful technician. To carry out the test, a *slump cone* with British Standard dimensions, is filled with concrete. This is done in four equal layers, each layer being tamped twenty-five times, the total time taken being not more than two minutes. The excess concrete is struck off the top and the slump cone mould removed. The amount of slump is measured. The real purpose of the test is to control reproducibility in successive batches of concrete. See figure 3.7.

Cube Test

This involves preparing concrete cubes of standard dimensions, followed, after curing in standard conditions, by the application of an increasing load, to determine the *crushing strength*. To obtain consistent and reproducible results, a standard procedure must be carried out. The concrete cubes are prepared in British Standard steel moulds, the concrete being placed in three equal layers, each layer being tamped a minimum of thirty-five times to ensure complete compaction. The top is trowelled flat. The moulds are stored for twenty-four hours at a temperature of 5–20 °C under damp sacking. The cubes are removed from the moulds and placed in water at 10–20 °C. The operations so far have been carried out on site, but the cubes may be removed to a laboratory for testing, wrapped in damp sacking.

To perform the test, a cube is placed between the platens of an upstroking or downstroking press, and force applied until crush-

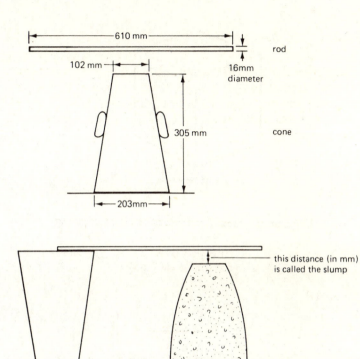

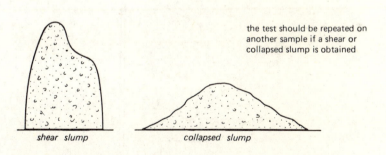

true slump

the test should be repeated on another sample if a shear or collapsed slump is obtained

shear slump   *collapsed slump*

Figure 3.7

ing occurs. The tests are normally carried out on seven-day and twenty-eight day old cubes. See figure 3.8.

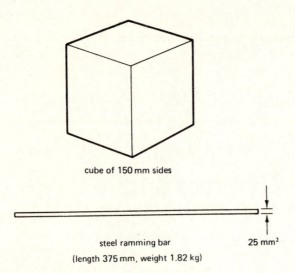

cube of 150 mm sides

steel ramming bar
(length 375 mm, weight 1.82 kg)

25 mm²

Figure 3.8

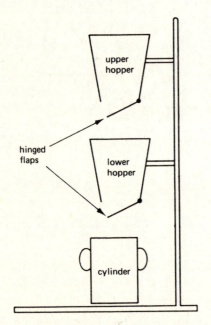

upper hopper

hinged flaps

lower hopper

cylinder

Figure 3.9

Compaction Factor Test

The results of this (figure 3.9) give a more accurate measure of the workability of fresh concrete, than the slump test. The principle involved is to allow concrete to fall freely into a container and measure the amount of compaction obtained.

Concrete is placed gently into the upper hopper until it is full. The bottom of the hopper is a hinged flap, which is opened letting the concrete fall into the lower hopper, where some compaction occurs. The flap of this hopper is opened next, and the concrete falls into a pre-weighed cylinder. Any excess concrete is removed from the top of the cylinder, which is weighed. This partially compacted concrete is fully compacted with a vibrator, and more concrete is added until the cylinder is full of fully compacted concrete. This is weighed.

$$\text{Compaction factor} = \frac{\text{Wt of cylinder full of partially compacted concrete} - \text{Wt of cylinder}}{\text{Wt of cylinder full of fully compacted concrete} - \text{Wt of cylinder}}$$

The compaction factor cannot be greater than 1. The closer it is to 1 the greater is the workability, because obviously a very workable and sloppy concrete will be almost compacted by the time it falls into the cylinder.

Compare the compaction factors in table 3.1.

Table 3.1

Use of Concrete	Compaction Factor	Workability
In concrete with complex re-inforcement	0.95	Very high
In simple structures not needing reinforcement	0.90	Medium
Reinforced concrete in large masses + vibration	0.85	Low
Large masses of concrete placed by vibration	0.75	Very low.

V—B Consistometer Test

This test (figure 3.10) is suitable for determining the workability of concretes which have low workability. It consists essentially of performing a slump test inside a metal cylinder, and measuring how long the slumped concrete takes by vibration to become completely compacted.

The slump cone is filled with concrete in four equal layers tamping each layer twenty-five times. The slump cone is removed and the transparent disc is placed so that it just touches the top of the slumped concrete. The vibrator and a stopwatch are both started together. The concrete is fully compacted when the transparent disc becomes completely coated with concrete. The time taken to do this is a measure of the workability of the concrete and is expressed as $V - B$ seconds. Obviously, the greater the workability the smaller is the number of seconds.

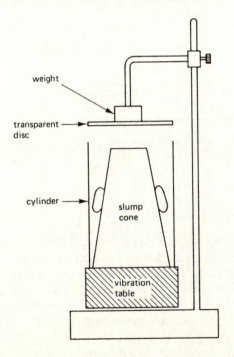

Figure 3.10

CURING OF CONCRETE

The production of concrete from cement is a chemical process

$$cement + water + aggregate = concrete + heat$$

The actual mechanism of the chemical change is very complex, but two points emerge readily from the equation.

(1) Since water is one of the reactants, it must be present for the curing process to take place. This is why it is important for the concrete to remain wet in the early stages—fresh concrete should be covered with damp sacking, or have water sprayed on it occasionally, or be covered with some impermeable sheeting such as plastics to prevent water evaporating. If the concrete is allowed to dry out too rapidly, shrinkage and/or loss of strength may result.

(2) The higher the temperature the faster the curing rate. In fact a 10 °C rise in temperature causes a doubling in rate of all chemical reactions. A temperature of 10–20 °C is very suitable for curing concrete. A very low temperature almost halts the curing, whereas too high a temperature can cause cracking due to thermal expansion, expecially in large masses of concrete.

Initial setting usually takes a few hours and curing needs several weeks, but the completion of the chemical process takes about a year.

Steam Curing

This method is used to accelerate the curing process and is useful for precast concrete. The steam is usually either at atmospheric pressure ('low' pressure) or at about 11 atmospheres ('high' pressure).

The high-pressure system achieves the most rapid curing times, and also gives a higher final strength than air-cured concrete. High-pressure curing gives concrete of twenty-eight-day strength in about thirty minutes, whereas 'low' or atmospheric pressure steam gives about half this strength in one day.

How Concrete gains Strength with Time

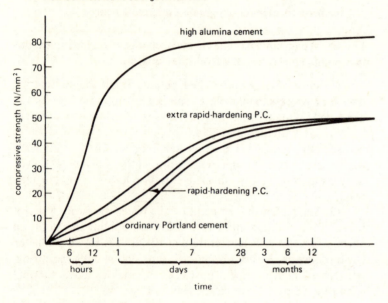

Figure 3.11

Notice in figure 3.11 that the most rapid gain of strength and the highest final strength are given by concrete composed of high alumina cement. Its use is restricted by high production costs.

All the concretes based on Portland cements achieve about 90 per cent of their final strength in twenty-eight days.

All concretes obtain their final strength after about one year.

How Concrete Hardens

How is it that by adding water to cement powder, the final product, concrete, is such a hard, rock-like substance? If water is added to sand alone, the same effect does not happen—the water

evaporates and the sand remains as granular and mechanically weak as ever.

When water is added to cement, some of the constituents of the cement dissolve in the water, producing eventually a *super-saturated* solution. These substances subsequently crystallise out to produce a large *interlocking, interpenetrating lattice* or *matrix* type of structure. It is this structure which gives concrete its strength.

The hardening process is completed by the absorption of carbon dioxide from the atmosphere.

Liberation of Heat during the Curing of Concrete (figure 3.12)

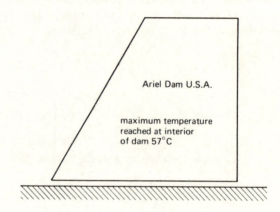

Figure 3.12

When concrete is poured in massive quantities, the amount of heat it liberates during hydration or curing is enormous. It was in connection with dams that the problem first received attention. The cooling process from the interior of a dam can take years, and it is the shrinkage or contraction accompanying this that causes cracking. The first dam made from low-heat cement was the Morris Dam, in the United States, built from 1932 to 1934.

As a very rough average, the amount of heat evolved during hydration is about *450 joules per gram of cement.*

CONCRETE ADDITIVES

Concrete is made basically from three materials—cement, aggregate and water. Sometimes, however, other materials are introduced into the mix, for the following reasons.

(1) To *retard* the setting of the concrete.
(2) To *accelerate* the setting of the concrete.
(3) To *waterproof* the concrete.
(4) To *introduce air* into the concrete.

Retarders

Gypsum is added to all Portland cements at the manufacturing stage (see p. 18) to prevent 'flash' setting. It operates by rendering insoluble certain substances present in cement. *Sugar* or *dextrin solutions* are added to surface concrete in precast units, so that later the concrete is brushed off to reveal and give an aggregate finished precast unit.

Accelerators

Calcium chloride and *sodium carbonate* solutions added to concrete mixes accelerate the setting of the concrete. This is important in jobs where time is limited.

Waterproofers

Soaps, sodium silicate (water glass), glues and *jellies* improve the waterproofing of concrete by reducing capillary attraction through the pores.

Air Entrainers

There are *two* distinct uses here.

(1) *Animal fats, vegetable fats* and *sulphonated oils* produce foaming in concrete on mixing. This entrains 3 to 6 per cent air in minute bubbles, evenly distributed in the concrete. These have the effect of improving workability by acting as a lubricant, and providing protection against frost by allowing water into the pores when it expands just before freezing.

(2) *Aluminium powder*, when added to a concrete mix, reacts chemically with alkaline components in the cement and produces bubbles of hydrogen gas. The amount of porosity produced by this means is much greater than 3 to 6 per cent and the aerated concrete produced, called *lightweight* concrete, has many desirable features (see below).

Lightweight Concrete

In the discussion on concrete mix design, it was stated as desirable that concrete should be as dense and non-porous as possible, and that 1 per cent of voids leads to a 5 per cent strength reduction. Also remember that the essential reason why the water content in a concrete mix should be kept to a minimum is to maximise concrete strength and density, and any excess water evaporates, leaving pores, and reduces strength.

However, if the structural strength and maximum density of concrete are not required, a highly porous concrete has many desirable properties such as lightness, good heat insulation, and so on. For this reason, much development work has been done on lightweight concrete, and it now has an important role to perform in the construction industry.

There are three principal ways of introducing pores or voids into concrete.

(1) Use a coarse instead of a graded aggregate (see p. 16) giving what is known as 'no fines' concrete.
(2) Use aerated aggregate (see p. 17), expanded pearlite for example
(3) Prepare a normal type concrete and aerate with aluminium powder or oils (see above).

A simple list of the relative merits and weaknesses of lightweight concrete follows. You can no doubt think of some more.

(1) *Light in weight*—easy to fix and work with; foundations can be less massive.
(2) *Good thermal insulation* due to air trapped in the pores.
(3) *Increased water penetration* because of capillary action in

the pores; thus exterior surfaces must be treated by rendering, etc.

(4) *Structually less strong*—unable to support heavy loadings.

(5) *Less raw materials needed*—the bulk density of lightweight concrete is less than that of the corresponding dense concrete.

EXERCISES

3.1 How can the properties of concrete be affected by
(a) porosity of hardened concrete
(b) grading of aggregate used
(c) moisture content of aggregate
(d) shape and strength of coarse aggregate particles?

3.2 State the main types of natural aggregate used for concrete making and describe where and how they are obtained and prepared.

3.3 Make a list of the essential qualities required in a good-quality aggregate.

3.4 How may the aggregate used influence the concrete produced from it?

3.5 What are the *four* main chemical compounds in ordinary Portland cement and how does each affect its properties?

3.6 Outline the general nature and use of the following cements compared with ordinary Portland cement
(a) extra rapid hardening Portland cement
(b) sulphate resistant Portland cement
(c) low heat Portland cement.

3.7 Outline the main stages in the production of cement clinker.

3.8 State *three* control tests, other than the compaction factor test, which are commonly used for checking the workability and strength of concrete. Describe in detail the compaction factor test.

3.9 Indicate briefly the stages in the setting and hardening of concrete.

3.10 Describe *four* main factors you would take into account and that would enable you to make efficient concrete.

3.11 What is meant by the workability of concrete? Explain concisely how this property can be measured.

3.12 Give a concise account of *three* types of lightweight concrete used in general building work or block-making. Include their advantages and disadvantages compared with dense concrete, and an outline of manufacture, general properties, characteristics and uses.

3.13 Describe how you would expect the quality of concrete to be affected by
(a) water/cement ratio
(b) cement/aggregate ratio
(c) methods of proportioning by weight or volume
(d) conditions of the aggregates
(e) the weather during mixing.

3.14 Indicate the important aspects of concrete mix design, and specify methods used for quality control with concrete. Include in your answer the technique for producing good concrete, and explain how its quality is affected by such factors as weather conditions, water content and cement content.

3.15 A concrete mix is to be batched by weight and the details of the quantities involved are as follows.

Material	Weight (kg)	Specific Gravity
Coarse aggregate	500	2.56
Fine aggregate	250	2.62
Cement	125	3.10
Water	40	1.00

The fine aggregate has a moisture content of 5 per cent calculated as a percentage of the wet weight and is included in the 250 kg. Calculate the following

(a) the actual weight of water present in the concrete mix

(b) the actual weight of fine aggregate used

(c) the yield in cubic metres if the concrete is thoroughly mixed and 100 per cent compaction is obtained

(d) the density of the wet concrete

(e) the density of the dry concrete, if a 0.25 water/cement ratio is necessary for complete hydration, assuming that all the excess water evaporates.

4 Water

FLOW OF WATER (HYDRODYNAMICS)

This is a very brief introduction to the flow of water through pipes and channels. Water is delivered to buildings in millions of litres every day and is usually an essential service. Obviously, pipes of adequate size must be used and gauges are needed for metering or measuring the flow. Open channels and gutters are also required to carry away rainwater, etc., and these must also be properly designed.

Streamline Flow and Turbulent Flow

It is a help to know, when designing pipes and channels, whether the flow is *streamline* (otherwise called laminar) or *turbulent*. The main features of these types of flow are given in figure 4.1.

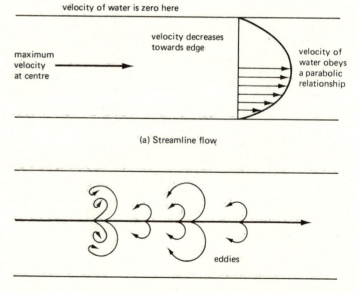

Figure 4.1

In practice, pumps are required to maintain the flow through pipes because collisions between water molecules, and with the sides of the pipe, are occurring constantly, and heat is being generated owing to these frictional effects. This leakage or wastage of energy causes velocity reductions in the water.

Generally streamline flow occurs in pipes, but if the velocity of the water is high or the pipe has a large diameter, turbulent flow usually occurs.

Bernoulli's Theorem

Bernoulli's theorem is really a special case of the *principle of conservation of energy*, which states that energy can neither be created nor destroyed, but may be converted from one form into another.

As water flows, the total energy possessed by its molecules is of three types, as indicated in figure 4.2.

Kinetic energy due to the velocity of the molecules.
Potential energy due to the position of the molecules.
Pressure energy due to the depth of the molecules in the liquid.

Bernoulli's theorem states that, provided there is no leakage of energy due to frictional effects, the total energy possessed by a moving liquid is constant, although the distribution of kinetic, potential and pressure energies among the molecules may be changing continuously. That is

$$\frac{kinetic}{energy} + \frac{potential}{energy} + \frac{pressure}{energy} = constant$$

In other words, as the water flows, the molecules in it are in constant random motion and colliding with each other in accordance with the kinetic theory. The energy of the constituent molecules is always changing, although the total energy remains the same. For instance, a molecule in contact with the side of the pipe may be knocked into the centre of the pipe, where the liquid velocity is a maximum; thus its kinetic energy increases. Another molecule may change its position and lose potential energy. However, the total energy does not change.

Venturimeter (figure 4.3)

It is useful to know the rate of flow of water through a pipe. The commonest method for doing this uses Bernoulli's theorem, and the technique used in a venturimeter is to place a constriction in the pipe, and measure the pressure difference with a manometer.

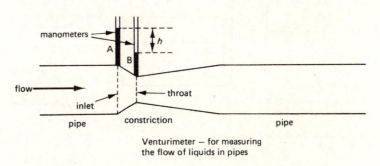

Venturimeter — for measuring the flow of liquids in pipes

Figure 4.3

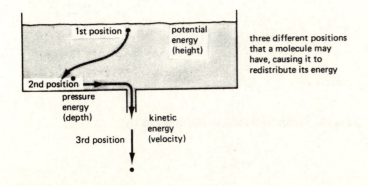

Figure 4.2

The height *h* is proportional to the velocity of water in the pipe. At the water pumping station, the height *h* is registered on a meter which is calibrated to give directly the rate of flow.

It can be shown that

$$Q = C_d A_1 \sqrt{\frac{2gh}{[(A_1/A_2)^2 - 1]}}$$

where C_d = discharge coefficient, a constant relating to the particular pipe ≈ 0.95, A_1 = inlet cross-sectional area (m²), A_2 = throat cross-sectional area (m²), Q = rate of flow (m³/s), g = acceleration due to gravity = 9.81m/s^2 and h = difference in height (m).

It is perhaps a surprise that the level in tube B is lower than in tube A. At the inlet the water possesses kinetic energy, potential energy and pressure energy; it is pressure energy that causes the rise in the manometer. At the throat (a constriction) the water flows faster, giving an increase in the kinetic energy. Since the potential energy is the same all along the pipe because it is horizontal, then at the throat, since the total energy is constant, the pressure energy must be reduced. Hence the level in manometer B is lower than in A.

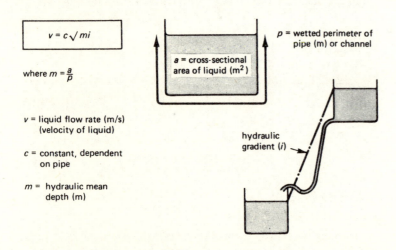

where $m = \frac{a}{p}$

v = liquid flow rate (m/s) (velocity of liquid)

c = constant, dependent on pipe

m = hydraulic mean depth (m)

Figure 4.4

Flow of Water in Channels and Sloping Pipes

When water is flowing horizontally through pipes, or uphill from, say, a pumping station to a reservoir, external pumping mechanisms are used. These pump the water and also overcome frictional effects within the liquid.

When water is flowing downhill, it falls under gravity and needs no external pump (figure 4.4). This phenomenon has been investigated by several people, the work and formulae of Chezy being the most acceptable.

Example using Chezy's Formula (figure 4.5)

A tank is full of water and has dimensions 3 m × 3 m × 2 m. Water flows from the tank down an open channel, 500 mm × 500 mm, inclined at a gradient of 1 in 200. The channel is one-third full of water. How long does it take to empty the tank ($c = 60$)?

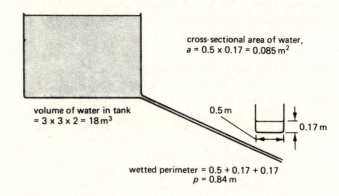

Figure 4.5

Solution Mean hydraulic depth is given by

$$m = \frac{a}{p} = \frac{0.085}{0.84} = 0.1 \text{ m}$$

Velocity of liquid down channel $= c\sqrt{mi}$

$$= 60 \sqrt{(0.1 \times 0.005)}$$
$$= 60 \sqrt{0.0005}$$
$$= 0.132 \text{ m/s}$$

Therefore

$$\text{volume flowing down channel} = 0.5 \times 0.17 \times 0.132$$
$$= 0.0112 \text{m}^3/\text{s}.$$

So

$$\text{time taken to empty tank} = \frac{18}{0.0112}$$
$$= 1605 \text{ s}$$

WATER SOFTENING

Water is obviously a very useful liquid for washing purposes. Owing to the attraction between water molecules, water tends to form a sort of skin over its surface, which reduces its washing ability. This surface effect is known as *surface tension*.

If the attraction between molecules is reduced, say by adding soap (figure 4.6), the 'skin' effect is lessened and the water performs its washing functions much more efficiently.

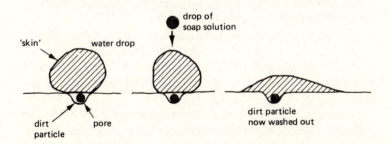

Figure 4.6

When water contains salts of the metals calcium and magnesium, it is said to be *hard* because these salts react chemically with soap or detergent, destroying its washing ability. Water that does not contain calcium and magnesium is said to be *soft*.

Water softening is therefore the process of removing calcium and magnesium salts from water, and rendering it suitable for washing and cleaning purposes. Most feasible processes have a chemical basis.

The commonest hardness-producing salts are calcium bicarbonate, $Ca(HCO_3)_2$, magnesium bicarbonate, $Mg(HCO_3)_2$, calcium sulphate, $CaSO_4$, and magnesium sulphate $MgSO_4$.

Methods of Softening Water

Boiling

When water is boiled, the bicarbonates change chemically into carbonates which are insoluble in water. These carbonates are thrown out, and settle as a whitish deposit known as lime, fur or scale. The over-all effect is their removal from the water. The calcium and magnesium sulphates are still present, so the water is still partially hard. This process is obviously not viable on a large scale.

calcium bicarbonate + heat → calcium carbonate (fur)
$$Ca(HCO_3)_2 = CaCO_3 + CO_2 + H_2O$$

magnesium bicarbonate + heat → magnesium carbonate (fur)
$$Mg(HCO_3)_2 = MgCO_3 + CO_2 + H_2O$$

Since the bicarbonates are removed simply by boiling, they give rise to what is known as *temporary hardness*, whereas the sulphates, which are more difficult to remove, give rise to *permanent hardness*.

Lime–Soda Process

This process, used at waterworks, consists simply of adding *lime* (calcium hydroxide) and *soda* (sodium carbonate) to the water.

The amounts of lime and soda added are carefully worked out, especially the lime, since an excess may possibly render the water harder than it was initially.

calcium bicarbonate + calcium hydroxide (lime) = calcium carbonate (fur) + carbon dioxide + water

$$Ca(HCO_3)_2 + Ca(OH)_2 = 2CaCO_3 + 2H_2O$$
$$Mg(HCO_3)_2 + Ca(OH)_2 = MgCO_3 + CaCO_3 + 2H_2O$$

calcium sulphate + sodium carbonate = sodium sulphate + calcium carbonate

$$CaSO_4 + Na_2CO_3 = Na_2SO_4 + CaCO_3$$
$$MgSO_4 + Na_2CO_3 = Na_2SO_4 + MgCO_3$$

Soaps and Detergents

The commonest soaps are the chemical compounds sodium stearate and sodium palmitate. They react chemically with hardness-producing substances in water, producing an insoluble compound—generally called scum! For example

sodium stearate (soap) + calcium sulphate → calcium stearate (scum) + sodium sulphate

When the calcium and magnesium ions have come out of solution, the water becomes soft.

Soap is made by heating together fat and alkali; for instance, dripping and caustic soda.

Chemically, soaps are salts of a group of acids known as fatty acids, two common ones being palmitic acid and stearic acid. They are all essentially long molecule substances, palmitic acid giving a soap *sodium palmitate*, $C_{15}H_{31}COONa$ (figure 4.7). This formula could be written

$$CH_3CH_2CH_2CH_2CH_2CH_2CH_2CH_2CH_2CH_2CH_2CH_2CH_2$$
$$CH_2CH_2COONa$$

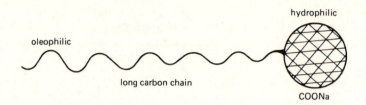

Figure 4.7

The washing mechanism of soap not only uses the principle of reducing the surface tension or 'skin' effect of water, but also the long carbon chain is soluble in fats and oils (oleophilic), whereas the head is water soluble (hydrophilic).

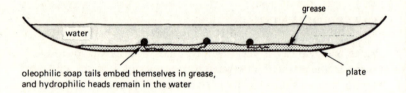

Figure 4.8

When the water is agitated, the soap molecules are removed, and in so doing, the oleophilic soap tails drag the grease off the plate, thereby cleaning it (figure 4.8).

As we said above, soaps are made by the alkaline hydrolysis of fats and oils, as shown in figure 4.9. These large molecules, called *triglycerides* split at the dashed line on heating with alkali, giving soap. Triglycerides are the basic constituents of fats and oils.

Detergents, which are potassium and sodium salts of a different group of acids, the *sulphonic acids*, are relatively new, scientifically based washing materials, and have wide applications.

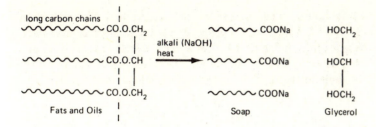

Figure 4.9

Ion Exchange Methods

The calcium and magnesium present in hard water are in their ionic form (see chapter 2). Ion exchange is the term given for a procedure which replaces Ca^{2+} and Mg^{2+} by other non-hardness producing ions, such as sodium, Na^+. The technique illustrated in figure 4.10 is a simplified version of a method introduced by the Permutit Company, and is usually called the *Permutit Process*.

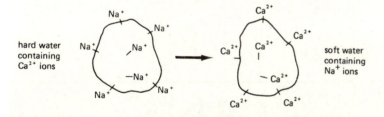

Figure 4.10

Hard water is passed over a resin containing sodium ions attached to its surface. Ion exchanges take place and the water is softened. The method is used in industry and commerce on a large scale. After a time, the resin no longer functions, because all of the Na^+ ions have been replaced by Ca^{2+} ions, but it can be regenerated by passing brine (sodium chloride solution) over the resin, when it becomes coated with sodium ions again and is ready for further use.

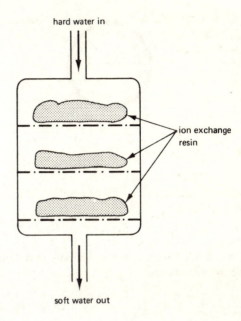

Figure 4.11

Merits of Soft Water

Soft water saves soap and detergent, so for industries which involve many washing processes, water softening is an economic proposition.

Merits of Hard Water

Although hard water causes 'scaling' or 'furring up' in boilers and pipes, a lining of scale (figure 4.12) does much to prevent lead poisoning by forming a protective lining. Also, since calcium is an essential element in bone formation and maintenance in humans, its presence in water is a useful source.

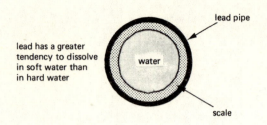

lead pipe

lead has a greater
tendency to dissolve
in soft water than
in hard water

water

scale

Figure 4.12

4.7 A house owner in Bristol has complained of a dwindling supply of hot water in his property. Describe fully the most likely cause of this, including how the water is affected in this way, how to measure its severity, and other effects this has. Also describe an economic method of removing the trouble from the water.

EXERCISES

4.1 Write down Bernoulli's equation for steady flow in an incompressible fluid, explaining the symbols and units usually used in their measurement.

4.2 Draw a labelled diagram of a venturimeter and explain its function.

4.3 Determine the rate of flow of water through a channel 1.2 m wide by 0.4 m deep. The gradient is 1 in 50 and there is a constant depth of flow of 0.3 m ($c = 80$).

4.4 Determine how long it would take to fill a bath $12 \text{ m} \times 6 \text{ m} \times 4 \text{ m}$ deep, if water flowed into it from a channel of rectangular section 400 mm $\times$ 200 mm deep, if the constant depth of flowing water was 160 mm and gradient of channel 1 in 50 ($c = 80$).

4.5 Give a precise account of the cause and effects of permanent and temporary hardness of water, and an economical method of removing both from a domestic water-supply system.

4.6 If you were in a hard water district state and explain
    (a) the effect you would expect on hot and cold water supplies and to be expected on supply systems
    (b) an economical method of softening the water.

5 Paints

Paint is an extremely useful material applied to protect surfaces and to make them attractive. Surface coatings are found on walls, furniture, cars, pencils, boilers, and so on, and their uses are legion. Paint can be applied to virtually any material: metal, wood masonry, brickwork, plastics, concrete or rubber.

Almost any shade or colour of paint can be produced, in a gloss, semi-gloss or flat/matt finish. Fluorescent paints are available which emit light. Spray paints are available which give a multicoloured spotted finish; there are paints which do not drip, and paints which kill insects settling on their surfaces.

COMPONENTS OF PAINT

Paints consist basically of four components (figure 5.1).

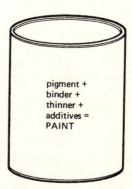

pigment +
binder +
thinner +
additives =
PAINT

Figure 5.1

Pigment: powder to give a paint *colour* and *opacity* (hiding power).

Binder: oil or resin to give a paint a final tough, coherent finish.

Thinner: liquid to reduce its viscosity ready for application.

Additives: a variety of these is added for a number of purposes (see p. 37).

Pigments

Pigments are put into paint to colour it, give it opacity or covering power, and usually improve its durability. Pigments should preferably be cheap, resistant to chemical attack, and light-fast (that is, not fade in sunlight). Most pigments are inorganic, since these are more stable than organic substances.

The best white pigment is titanium dioxide, which has very good opacity and durability. Others are white lead and zinc oxide. Nearly all black pigments are forms of carbon. With coloured pigments, the commonest reds and browns contain iron oxide, while more brightly coloured pigments are compounds of various metals such as chromium and cadmium (for example, cadmium yellow). Colouring is also achieved with various organic dyes. Metallic finishes are achieved by incorporating aluminium, copper and other metallic powders.

It is perhaps just as well to mention in this context a group of powders called *extenders*, which are usually white substances such as barium carbonate, talc and silica. These materials impart very little colour or opacity to a paint, and are put in to cheapen the paint, improve its brushing characteristics and toughness and reduce gloss.

Binders

The binder, or medium, as it is sometimes called, forms part of the *vehicle* of a paint (figure 5.2). The thinner and additives also form the vehicle.

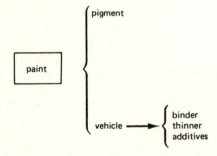

Figure 5.2

There are three types of medium, or binder: oils, resins and others.

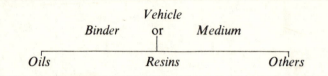

Oils	Resins	Others
Generally of vegetable origin; classified in chemical terms as triglycerides (see p. 32); they are all *drying* or *unsaturated* oils, e.g. linseed oil, tobacco seed oil, coconut oil, tung oil, soya bean oil, castor oil, 'blown' linseed oil	Very broad range of substances including everything used as a binder in paint except oils and 'others'; may be naturally occurring or synthetically produced, e.g. *natural*—congo copal, dammar resin, shellac—*synthetic*—alkyd, polyester, acrylic, vinyl, styrene and polyurethane	Miscellaneous media, e.g. glue, gelatin and starch

A binder or medium is a substance which dries by evaporation, oxidation or polymerisation, to give the paint a hard durable surface, and to bind the pigment together.

The drying of oils and resins is a complex chemical process, but in simple terms it involves unsaturated bonds between atoms, usually carbon. (Double and triple bonds are said to be unsaturated.)

single bond (saturated)   *double* bond (unsaturated)   *triple* bond (unsaturated)

The double and triple bonds, being chemically unstable, break up, and the free linkages formed join on to further atoms, making giant molecules or polymers (see p. 1) as in plastics. Sometimes oxygen bridges are involved.

$$\equiv C - O - C \equiv \qquad \overset{\displaystyle O}{- C \triangle C -}$$

This process of polymerisation and oxidation leads to the formation of giant cross-linked three-dimensional molecules, which cause the formation of the hard skin characteristic of paint.

Modern oil-based paints generally contain a binder which is a mixture of oil and resin. When the paint is brushed out, the large surface area exposed gives access for oxygen in the air to cause oxidation, and with simultaneous polymerisation, the paint dries out. The term 'dries out' is something of a misnomer, because although evaporation of thinner takes place, the 'drying' is a chemical hardening process.

Some paints, which dry by polymerisation, will only do so at high temperatures. This means that the painted object has to be 'stoved' or placed in an oven for the paint to dry. Although this is laborious and costly, the finished paint surface is usually extremely hard and durable.

Emulsion paints, which are very popular in house decoration, dry by evaporation. The binder or medium in these paints consists of a suspension of polymer particles in water. These media are prepared usually by the emulsion polymerisation process described on p. 4. Polyvinyl acetate emulsion is often used, or copolymers made by polymerising two monomers together, such as vinyl acetate and butyl acrylate.

In the past, simple whitewashes of chalk and water bound with starch have been used, but they are only used for temporary decoration.

Mention should perhaps be made of 'blown' linseed oil, which is prepared by blowing air through linseed oil, causing it to undergo partial oxidation and polymerisation. It is then incorporated into paint, giving it quick-drying properties.

Thinners

These are liquids which are mixed into paint to reduce its viscosity, increase its flowing properties and prepare it for application to a surface. After the paint has been applied, the thinner rapidly evaporates. Thinners commonly used include water, turpentine, white spirit and ethyl acetate.

Additives

Driers

These are catalysts and oxidising agents introduced to speed up the drying process, by providing oxygen and/or increasing the rate of polymerisation; for example benzoyl peroxide and cobalt napthenate.

Insecticides and Fungicides

These are introduced to kill flies and so on, and to prevent mould growth; they include aldrin, dieldrin and phenol derivatives.

Light Stabilisers

Some paints, particularly white and lightly coloured ones, yellow with ageing. This is due to exposure to sunlight, which tends to break down the polymeric structure of the binder, setting up carbon polymer chains containing alternate single and double bonds (known as a *conjugated* system) which look like

$$\equiv C - C \equiv C - C \equiv C - C \equiv C - C \equiv C -$$

This type of structure always produces an orange or yellow colour. It is, for instance, present in carrots in the form of a substance called *carotene*. Stabilisers which deal with this problem include dibutyl tin dilaurate.

Inhibitors

These substances are introduced into some paints to slow the drying process. For instance, hydroquinon decreases the polymerisation rate of synthetic resins.

Antisettling agents

These are put into paints to retard sedimentation of the pigment and increase shelf-life.

Bodying agents

These are added to increase the viscosity of a paint and improve its few properties, for example methyl cellulose. A recent development is the use of polyamides to produce *thixotropic* paints, which have the unique property of losing viscosity on being stirred, but thicken up again once the stirring ceases (non-drip paints are of this kind).

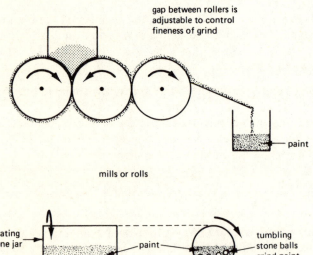

Figure 5.3

PAINT MANUFACTURE

First a formulation is drawn up for the paint; that is, it is decided which ingredients are to be used, and their proportions. The pigment, and any extender used, is mixed with some of the binder and ground up until the appropriate particle size is reached. This is usually done either on rolls or by ball-milling (figure 5.3). Meanwhile the additives are mixed in with the rest of the binder. The two parts are then mixed together to give the paint, which is reduced in viscosity with thinner to the specification required. After testing it is put into tins and despatched.

TYPES OF PAINT

There is a bewildering variety of paints available nowadays. The following is really a brief catalogue of the principal ones.

(1) *Primer*—applied to a bare surface or substrate to protect it and provide a good surface for subsequent layers of paint (that is, undercoat and top coat). Primers are available for wood and metal. Metal primers often contain lead compounds or zinc chromate.

(2) *Undercoat and top gloss finish*—formulations for these two types of paint differ in that gloss paint has a higher binder/pigment ratio than undercoat, which is matt or flat (figure 5.4).

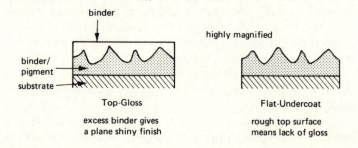

Figure 5.4

(3) *Aluminium paint*—contains aluminium powder and gives good *heat resistance*.

(4) *Bituminous paint*—contains a tarry base and gives good *water resistance*.

(5) *Emulsion paint* (see also p. 37)—very good for indoor decoration; its advantages are that it is thinned with water, is easy to apply, dries quickly, and can be obtained in almost any colour.

(6) *Fungicidal/pesticidal paint*—contains an appropriate pesticide or insecticide, such as DDT.

(7) *Fire-resistant paint*—contains an asbestos filler.

(8) *Lacquer*—surface coating in which the polymerised binder dissolves in the solvent or thinner, for example, cellulose nitrate lacquer. The feature of these paints is that they dry very rapidly. They are used extensively in the car industry.

(9) *Rubber paint and chlorinated rubber paint*—have very good *chemical resistance* and hence are used where other paints would be inappropriate, for instance, in chemical works and breweries.

(10) *Stoving enamel*—dries by polymerisation only at elevated temperatures, usually between 60° and 160°C (see p. 37).

(11) *Thixotropic paint*—non-drip paint.

(12) *Varnish*—effectively paint without any pigment; it is normally made by mixing oil, resin and drier; such a clear material is useful for floors and furniture.

TESTING PAINT

Many tests are available for the raw materials of paint: tests on pigments, oils, and so on, but we shall restrict ourselves here to tests on the paint itself. Again many tests are available, with British Standards specification, but we shall only deal in principle with the most important ones.

When the paint has been produced in the works, thinner is added to produce the correct *viscosity*. To calculate how much thinner is required, laboratory tests are carried out, and the simplest method is to use a Ford cup, figure 5.5. Sufficient thinner is added until a given volume of paint flows through a small orifice in a given time, and at a given temperature. More elaborate

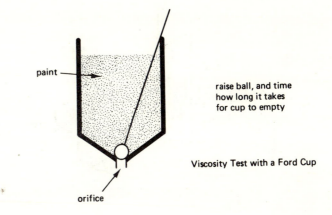

Figure 5.5

equipment, 'viscometers', is also available for measuring the viscosity of paint.

At this stage pigment *particle size* can be checked by mechanical means, and the polymer particles of an emulsion for emulsion paint can be checked with a microscope.

If the paint is coloured, it must be colour-matched with the previous batch of paint of the same colour. This is done by adding dyes and blending paints, and much depends on the skill of the operator. *Reflectance* tests can be performed by allowing light to fall on to dried films of the paint, measuring the amount of reflected light with a photocell or lightmeter.

The *drying time* of the paint can be measured in a variety of ways; for instance, by allowing a needle to move through the paint on a slowly moving panel over several hours; or by allowing sand to fall on to a panel of paint while it moves.

Hardness tests may be performed on the dried paint film.

An important long-term test is to observe how the paint *weathers*. Painted panels are prepared, and left in exposed places to see what effect sunshine, rain, snow, and so on, have on them. Several paint panels are prepared, and they are exposed in different locations—particularly in industrial atmospheres, and in marine environments.

The weathering process obviously takes years to complete,

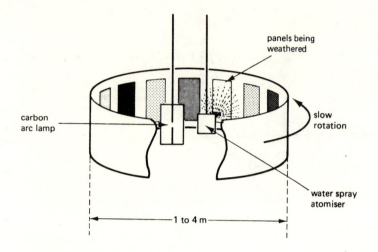

Figure 5.6

which is a serious disadantage when you want to market a new paint. Therefore a technique known as accelerated weathering has been developed (figure 5.6), in which rainfall is simulated by water sprays, and powerful carbon-arc lamps direct intense light on to the paint films. In this way, several years' weathering can be achieved in a few weeks.

DEFECTS IN PAINT

After the paint film has been applied and has dried out, various factors can spoil its final appearance. The most important ones are given opposite, together with a cause and a remedy.

Defect	Cause	Remedy
Bleeding: discoloration due to dark substances, especially tar, and resin from wood knots	Tar from, say, chimney breasts, and knot resins, bleed up to the paint surface	Apply aluminium foil or some other impermeable coating, and repaint
Blistering: bubbles or blisters on paint surfaces	Damp substrate. Water tries to evaporate under the paint surface	Apply paint only to dry substrates
Bloom: appearance on paint surface like dust	'Bloom' is ammonium sulphate and its formation is due to atmospheric conditions	Wipe off with a cloth
Chalking: on rubbing the paint surface, pigment comes away	Lack of binder, either due to age, or faulty paint formulation	Clean off, and apply new paint
Cissing or pinholing: paint retracts from certain areas on substrate	Incompatible foreign matter on substrate	Clean substrate thoroughly before applying paint
Curtaining: paint runs, giving an undulating appearance	Paint film applied too thickly	Apply paint in thinner coats
Orange peeling: surface appearance like an orange	Due to faulty paint-spraying technique	Apply with correct technique
Peeling or flaking: paint peels or flakes off the surface	Damp substrate	Apply paint only to dry substrates

EXERCISES

5.1 What is meant by the following
    (a) opacity
    (b) thixotropy
    (c) drying oil
    (d) viscosity
    (e) light fastness
    (f) vehicle
    (g) 'blown' linseed oil?

5.2 Suggest how to cure the following defects in paint
    (a) blistering
    (b) flaking
    (c) chalking
    (d) cissing.

5.3 Write down three facts for each example which would be relevant in formulating the following
    (a) a high-quality white gloss finish coat
    (b) a cheap surface coating for an outhouse
    (c) a quick-drying flat black undercoat
    (d) a wood primer
    (e) a low viscosity grey undercoat with good opacity.

5.4 Give four advantages of oil paints over emulsion paints.

5.5 What is meant by the following
    (a) accelerated weathering
    (b) shelf-life of paint
    (c) ball-milling
    (d) drying time
    (e) colour matching
    (f) stoving enamel
    (g) volatile thinner
    (h) substrate?

5.6 (a) List the properties expected of a good oil paint.
    (b) State the principal ingredients and their function.

(c) Describe briefly the cause, appearance and remedy for two defects which occur in paintwork.

5.7 Write down the cause, effect and cure of *six* defects found in paintwork.

5.8 Write a concise account on emulsion paints.

5.9 Suggest a simple method of comparing the viscosities of oils to be used in paint manufacture, so that they may be placed in order of increasing viscosity.

6 Stones and Ceramics

STONES

Natural stones fall into three groups: *igneous rocks*, *sedimentary rocks* and *metamorphic rocks*.

Igneous rocks were formed by the solidification of molten material, and as such are generally non-porous; *granites* are a typical example.

Sedimentary rocks are those which have been formed by a sedimentation or settling process—usually on the sea bed. *Sandstones* and *limestones* were both produced in this manner. Sandstone, as its name suggests, is largely silica or quartz (forms of silicon dioxide). Limestone, on the other hand, is mainly calcium carbonate. Some limestones also contain magnesium carbonate, and the resulting mixture is called dolomite; the Dolomite Mountains in Europe contain this magnesium carbonate/calcium carbonate mixture. A disadvantage of some sandstone and limestone, when used as building stone, is that it contains hard and soft layers (being sedimentary) which weather at different rates, and this gives a poor appearance.

Metamorphic rocks have been formed from sedimentary rocks which at some stage have been subject to heat and pressure and become molten. At a later stage, resolidification has led to the creation of what is known as metamorphic rock. *Slates* were formed when clay resolidified, and *marbles* formed from limestone treated in the same manner.

Properties of Stones

Density

As we would expect, the sedimentary rocks are the least dense, being porous. Compare the densities in table 6.1 with that of water (1000 kg/m^3).

Strength

As with concrete, the resistance a stone offers to crushing is used as a measure of its strength. Again, like concrete, the strongest stones are those which are least porous. Notice that the strength of stone

Table 6.1

Stone	Density (kg/m^3)
Granites	2600–3200
Sandstones	2100–2800
Limestones	1950–2400
Slates	2800–3100
Marbles	2900–3000

compares very favourably with that of concrete (see table 6.2). Stones are also similar to concrete in being strong in compression and weak in tension.

Table 6.2

Stone	Strength (N/mm^2)
Concrete	20–70
Granite	100–300
Sandstone	20–170
Limestone	15–45
Marble	200–250

Effect of Heat

All stones tend to crack and spall when subjected to high temperatures. This is because they are poor conductors of heat, although their coefficients of expansion are not high.

Effect of Water

The amount of water absorption experienced by a stone depends on its porosity; granites, for example, being non-porous, absorb no water at all. Limestones attract much water by capillary action through the pores.

When water freezes, it expands, due to structural changes at molecular level. If this happens inside damp stonework, damage occurs.

Another cause of damage is due to absorption of water containing soluble salts dissolved in it. In dry weather, the water evoporates and the salts crystallise out inside the stone, setting up stresses.

Acid Attack

On the whole, stones are extremely durable. However, some damage can be caused, particularly to limestones, by carbonic and sulphurous acids. Carbonic acid forms when carbon dioxide dissolves in rainwater. Sulphurous acid forms when sulphur dioxide, present in polluted atmospheres, dissolves in water.

Tests for Stones

Since stones occur naturally, and cannot be 'tailor-made' to suit a particular job, like materials such as concrete, paint, and so on, tests are usually performed to check their suitability. The list below indicates some of the tests that are carried out. Many of them have procedures laid down by the British Standards Institution.

(1) *Microscopic examination* for stratification, pores, cracks, and non-homogeneity.
(2) *Crushing strength test.*
(3) *Porosity and water absorption* tests.
(4) *Freezing* test: the stone is repeatedly frozen and unfrozen and examined for frost damage.
(5) *Acid* test: the stone is placed in acid and subsequently examined.
(6) *Sodium sulphate crystallisation* test: the stone is placed in sodium sulphate solution, and allowed to dry out.

Investigations are conducted for damage caused by the extra volume of the salt within the stone.

Defects occurring in Stonework

Although most of the defects occurring in stonework have already

Table 6.3

Defect	Cause	Remedy
Cracks in granite	Thermal expansion movement	Replace cracked stones
Efflorescence in porous stones—limestones and sandstones; that is, a white saltlike substance on the face of the stone	Crystallisation of soluble salts (see also p. 46)	Wash down the face of the stone
Uneven ridge-like-appearance in limestones and sandstones	Due to erosion of soft layers in the sedimentary stone	Replace damaged stones; good-quality stones are not subject to this defect
Cracking and spalling in limestones and sandstones	Due to frost; porous sedimentary stones absorb water, which expands on freezing, and causes cracking	Replace damaged stones; lay the stones with their strata horizontal
Formation of a glassy skin of calcium sulphate/sulphite on the face of the stone; this breaks off causing damage	In polluted atmospheres oxides of sulphur form acids with water, attacking limestones, dolomites and calcareous sandstones	Clean the face of the stone by washing, or sometimes sandblasting
Staining and cracking on most stones	Corrosion of embedded metal in stones, leading to stains and salt crystallisation	Treat the metal before embedding in the stone

been mentioned, table 6.3 summarises the principal ones, and offers some methods of prevention and cure.

CERAMICS

A ceramic is generally defined as any product which is made from burnt clay.

Clay

Soils, sands, marls, clays and silts are chemically similar, all resulting from the erosion of rocks, and containing oxygen and metals such as calcium, silicon, iron, aluminium, magnesium, sodium and potassium.

The most common clay is *feldspar*, which is a mixture of two complex substances: $K_2O.Al_2O_3.6SiO_2$ and $Na_2O.Al_2O_3.6SiO_2$. On erosion and weathering, both give another substance: $Al_2O_3.2SiO_2.2H_2O$. (Potassium oxide, K_2O, and sodium oxide, Na_2O, are both removed, being water soluble.)

Types of Clay

Terracotta Clays

These are widely distributed, and are used to make a great number and variety of ceramic products for use in the construction industry. Since these clays contain iron oxide, which is a red-brown colour, the products are similarly coloured. Bricks and tiles are produced in very large quantities.

Kaolins

These clays are white, so they are used for making bone china. They consist largely of $Al_2O_3.2SiO_2.2H_2O$. (Note that of all the metallic oxides present in clays, all are white or colourless apart from iron oxide. The amount of iron oxide present in a clay determines whether it is red, red-brown, or of a lighter shade. Since kaolins contain no iron oxide at all, they are white.) Kaolin is also a refractory clay, that is, it is heat resistant.

Stoneware Clays

These clays can become molten at lower temperatures than other clays, and are therefore used in vitrified or stoneware products.

Fire Clays

These clays, like kaolins, are refractory, but since they contain some iron oxide (1 to 2 per cent), they do not give pure white products. They are therefore unsuitable for use in high-quality crockery, but are eminently satisfactory for flue liners and for furnace refractory linings.

Ball Clays

These clays contain little or no iron oxide, and so tend to be white. However, their firing technique is complicated by the fact that some organic material is usually present, which burns off in the firing furnace, causing loss in weight.

Uses of Clay in the Construction Industry

After a clay has been selected for a particular product, it is moulded by extrusion or other processes, and fired by placing in an oven or furnace at an elevated temperature.

Products made include

(1) bricks
(2) roof, floor and wall tiles
(3) sanitary fittings
(4) refractory products (such as flue liners).

Firing Processes

The strength of ceramic products in the construction industry is all important. In the same way that cement, which is powdery and mechanically weak, can give rise to extremely hard concrete, so

clay can produce extremely hard bricks, tiles and so on. However, whereas with cement and concrete the hardening process is chemical by nature, with clay the hardening or firing process is more physical. The strength of the final ceramic product depends on the temperature to which it has been subjected.

Before firing, clay products are porous. Heating tends to melt them, and the material which resolidifies from the molten state is non-porous, and therefore harder and stronger. (Compare the strength of a thick glass with that of sand, from which it is made.) The strength of a fired clay product therefore depends on the degree of melting or *vitrification* which it has undergone, and this depends on the firing temperature. A brick which has been fired at a high temperature will be dense, relatively non-porous, and of a high compressive strength. Compare the figures in table 6.4.

Table 6.4

Type of Brick	Firing Temperature (°C)	Density (kg/m^3)	Porosity (per cent)	Compressive Strength (N/mm^2)
Common brick	950–1150	1200–1600	55–45	35
Facing brick	1000–1250	1900	30	40
Engineering brick	1200–1300	2500	5	70

The strongest bricks would, of course, be obtained by heating them until they were molten, and then allowing them to resolidify, but this is uneconomical and unnecessary.

The firing process involves a number of different stages. The wet clay extruded products are firstly allowed to dry, either in a special drier or in a special chamber in the firing kiln. They are then fired in the kiln or furnace.

At temperatures up to 150 °C, any excess water evaporates off. This is a purely mechanical process.

At temperatures between 400 and 600 °C, water is still coming off, but now it is being stripped from the molecules of clay, a typical formula being $Al_2O_3.2SiO_2.2H_2O$, where

2 $$Al_2O_3.2SiO_22H_2O \rightarrow Al_2O_3.2SiO_2 + 2H_2O$$

This is a process of chemical dehydration.

Any organic material is also being burnt off as carbon dioxide and water, up to temperatures of about 650 °C. Above this temperature, the vitrifying, or firing, or partial melting process takes place, to give the final strong product.

Efflorescence

This word has two meanings which are quite distinct (and is not to be confused at all with *effervescence*, which means fizzing or bubbling in liquids!) In *chemistry* it is the word used to describe the phenomenon shown by some crystalline substances of losing their water of crystallisation after being exposed to the atmosphere. For example, crystals of washing soda gradually fall to a white powder when they have been left in the open

$$Na_2CO_3.10H_2O \rightarrow Na_2CO_3H_2O + 9H_2O$$
crystals   powder

Now in *building*, it describes the presence of white salts which sometimes appear on the face of new brick walls. This is due to soluble salts in the clay from which the bricks were made; usually magnesium and sodium salts. When the wall becomes damp, water soaks into the brick and dissolves these salts. Later, in the drying process, the solution moves to the face of the brick and evaporates, leaving the salt behind as a white deposit. If unsightly, it can be washed off.

A method used by some brick companies to cure efflorescence is to mix a little barium carbonate into the clay. This renders the salts insoluble, so that they do not dissolve in any penetrating water and remain in the interior of the brick. For example

magnesium + barium → magnesium + barium
sulphate   caroonate   carbonate   sulphate

(soluble salt)   (both insoluble in water)

EXERCISES

6.1 What does *vitrification* mean, in brick-making?

6.2 What is a sedimentary rock?

6.3 Name a cure for efflorescence in brickwork, writing down the chemical reaction involved.

6.4 Why are *terracotta* clays red?

6.5 Why are *kaolins* white?

6.6 In brick firing, what do the terms 'mechanical dehydration' and 'chemical dehydration' mean?

6.7 Why does a higher firing temperature generally produce a stronger brick?

6.8 Why are the igneous rocks generally more dense than sedimentary rocks?

6.9 What is a refractory clay?

6.10 How was marble formed in the Earth's crust?

6.11 What is magnesian limestone?

6.12 How does cracking in granite take place?

6.13 (a) Assume you are given responsibility for the maintenance of several buildings constructed of calcareous and siliceous sandstone, limestone and granite in an industrial area. Indicate *five* defects you would expect to find in the stonework, including the cause, method of eradication, and prevention if possible.

(b) List *five* tests that could be applied to stones to assist in their selection for a particular contract.

7 Plasters

GYPSUM PLASTER (BS 1191)

All plasters are made of hydrated calcium sulphate, $CaSO_4.2H_2O$. This means that each molecule of calcium sulphate, $CaSO_4$, is bonded to two water molecules. When water is attached to molecules in this fashion, it is called water of crystallisation.

Two basic raw materials are used for making plaster: *gypsum*, $CaSO_4.2H_2O$ and *anhydrite*, $CaSO_4$.

When gypsum is heated, it loses three-quarters of its water and becomes *plaster of paris*

$$CaSO_4.2H_2O \xrightarrow{} CaSO_4.\tfrac{1}{2}H_2O + 1\tfrac{1}{2}H_2O$$
$$\text{gypsum} \qquad\qquad \text{plaster of paris}$$

When plaster hardens or sets, the chemical process involved is one of hydration. With all types of plaster, only two reactions are involved, which are

$$CaSO_4 + 2H_2O \rightarrow CaSO_4.2H_2O$$
$$CaSO_4.\tfrac{1}{2}H_2O + 1\tfrac{1}{2}H_2O + 1\tfrac{1}{2}H_2O \rightarrow CaSO_4.2H_2O$$

Water is added to the anhydrous calcium sulphate, $CaSO_4$, or to the hemihydrate $CaSO_4.\tfrac{1}{2}H_2O$. The calcium sulphate dissolves in the water, and soon forms a saturated or supersaturated solution, from which crystals of the dihydrate separate out. These crystals form a *strong interlocking matrix*, which is the reason for the setting and strength of plaster (compare this with the setting of cement, p. 24). Also, again like concrete, excess water causes porosity. The setting of plaster is therefore a crystallisation procedure.

Of the two hydration or setting processes

$$CaSO_4 + 2H_2O \rightarrow CaSO_4.2H_2O \tag{7.1}$$

gives *slow setting*

$$CaSO_4.\tfrac{1}{2}H_2O + 1\tfrac{1}{2}H_2O \rightarrow CaSO_4.2H_2O \tag{7.2}$$

gives *rapid setting*.

Table 7.1

Type	Class A	Class B	Class C	Class D
Name	Plaster of Paris	Retarded hemihydrate plaster	Anhydrous plaster	Keene's plaster (or cement)
Speed of set	Fast (very)	Fast	Slow, but continuous	Slow (very) but continuous
Basic raw material	Plaster of Paris $CaSO_4 \cdot \frac{1}{2}H_2O$ (7.2)	Plaster of Paris $CaSO_4 \cdot \frac{1}{2}H_2O$ (7.2)	Anhydrite $CaSO_4$ (7.1)	Anhydrite $CaSO_4$ (7.1)
Actual composition	Plaster of Paris	Plaster of Paris + retarder (e.g. keratin) to slow the speed of setting	Anhydrite + Accelerator (e.g. Alum) to increase the setting rate.	Anhydrite + Accelerator (e.g. alum) to increase setting rate
Uses	Since the set is so rapid, its use is restricted to plastering very small areas, stopping cracks and holes, and for casting in moulds	This is the most useful plaster; its setting time can be controlled by the amount of retarder added	Due to the slower setting time, it may be worked longer, and is thus very useful for finishing coats	Slow setting, but gives a very good hard finish, suitable for edges, and where hard surfaces are required, e.g. squash courts

TYPES OF PLASTER

Four types of plaster are available, as classified in BS 1191. These are described simply as Class A, Class B, Class C and Class D (see table 7.1).

When gypsum, $CaSO_4 \cdot 2H_2O$, is heated to about 170°C, plaster of Paris is formed, $CaSO_4 \cdot \frac{1}{2}H_2O$. The capacity of plaster of Paris to rehydrate back to gypsum is very great, and so set is rapid.

On the other hand when gypsum is heated above 200°C to remove all its water of crystallisation, and become anhydrite, $CaSO_4$, the capacity which anhydrite has for rehydration is limited, and so Class C and Class D plasters are slow setting. The anhydrite is said to be 'dead-burnt'. The anhydrite for use in Class C plaster is heated or calcined to about 250°C, and that for Class D plaster to about 360°C.

Workability Agents for Gypsum Plaster

Substances can be introduced into gypsum plaster to make it easier to work. There are two types: *slaked lime* and *organic plasticisers*.

Slaked Lime, that is, calcium hydroxide, $Ca(OH)_2$ acts by slowing the setting time, and by holding in water, since it has good water retention.

Organic plasticisers, which are obtainable in liquid and powder form, entrain tiny air bubbles, increasing the workability of the plaster, but not affecting its setting time.

Defects in Gypsum Plaster

Many faults can arise on surfaces composed of gypsum plaster. Some of the principal ones are as follows.

(1) *Bubbling* and *blistering* (otherwise called pitting, blowing and popping): caused by moisture coming into contact with

unslaked lime in the plaster, and by water reacting with powdery plaster left by inefficient mixing. In both cases, the bubbles and/or blisters arise from thermal expansion and water evaporation due to the heat evolved during hydration.

(2) *Cracks*: caused by a variety of things but principally because of settlement and mechanical impact (notice, for instance, how plaster cracks around door frames).

(3) *Efflorescence*: the process is similar to that occurring in brickwork (see p. 46). Soluble salts in plaster (and sometimes in the bricks or blocks behind it) dissolve in water, which later evaporates from the surface of the plaster, leaving a white salt-like appearance.

(4) *Non-adhesion*: there is a number of reasons why plaster sometimes does not adhere to a surface, the principal one being lack of surface area. Generally, the greater the surface area of the backing cost, the better is the adhesion.

(5) *Rust staining*: sometimes, calcium ions in the plaster change places with metal ions present, say, for example, in iron and lead pipes. This effectively causes corrosion of the metal pipes. If the metal ions liberated into the plaster are coloured, then colouring or staining occurs. This is prevented if the pipe is treated before embedding in the plaster.

LIME PLASTER

Limestone is mainly calcium carbonate, $CaCO_3$. On heating this material in a lime kiln, quicklime is produced. On adding water to quicklime, the product is calcium hydroxide, $Ca(OH)_2$ (slaked lime, or hydrated lime).

$$\underset{CaCO_3}{\text{calcium carbonate}} \xrightarrow{\text{heat}} \underset{CaO}{\text{calcium oxide}} \xrightarrow{\text{water}} \underset{Ca(OH)_2}{\text{calcium hydroxide}}$$

Slaked lime, or calcium hydroxide, is the basic raw material for lime plaster. After application, it hardens in two stages. First water evaporates from it, and any hardening is due solely to this effect. In time, over months and years, carbon dioxide in the atmosphere

reacts chemically with the slaked lime to give a much harder product. Its principal disadvantage, and the main reason why it is little used today, is the very lengthy setting time.

$$\underset{\text{soft}}{Ca(OH)_2} + CO_2 \xrightarrow{\text{long time}} \underset{\text{hard}}{CaCO_3 + H_2O}$$

Lime is often put into gypsum plaster, to improve its workability and retard its setting time.

Hydraulic Lime

Limestone and chalk, both forms of calcium carbonate, are usually white, whitish, or yellowish. Another form of calcium carbonate, called grey chalk or greystone, contains impurities, notably alumina and silica. When this material is heated in a kiln, calcium oxide or lime is produced, but silicon dioxide and aluminium oxide are also present in a complex mixture which resembles cement (see chapter 3). On adding water, it sets more quickly than ordinary lime. It is called hydraulic lime because it can set under water.

EXERCISES

7.1 Give a brief account of Class A and Class B gypsum plasters.

7.2 Report on the manufacturing and general characteristics of two types of gypsum plaster used in building.

7.3 Outline the cause, effect, remedy and method of preventing pitting and popping in plaster.

7.4 Give a concise account of the characteristics and uses of *two* types of lime used in building.

7.5 Describe the types, properties and characteristics of lime.

7.6 Explain the similarity between hydraulic lime and ordinary Portland cement.

8 Timber

Timber is the hard fibrous material obtained from trees. Owing to its strength, lightness, durability and wide-ranging availability, it is one of the world's most important materials, and is ideal for use in the construction industry.

Timber falls into two categories: *hardwood* and *softwood*. Softwood comes from trees which have needle-shaped leaves and are coniferous, such as pine and spruce. Hardwood comes from deciduous trees with broad leaves, such as oak and elm. Although most of the hardwood classified as hardwood by the above definition is in fact 'hard', and most classified softwoods are 'soft', there are exceptions; balsa wood, the world's softest wood, comes from a deciduous tree, and is therefore classed as a hardwood.

Softwoods are used for general constructional purposes in boards, beams, door frames, and so on, while hardwoods are used for making quality joinery, durable sills, and so on.

The trunk of a tree, from which timber is obtained, consists essentially of bundles of long thin cells or tubes made of cellulose, 'stuck' together with a material called lignin.

The tree grows by producing sugars by a process called photosynthesis. The raw materials needed for this process are carbon dioxide, absorbed from the atmosphere, and water (plus salts) which is absorbed at the roots of the tree. Photosynthesis takes place in the leaves of the tree, and water proceeds from roots to leaves via the hollow thin tubes in the sapwood.

$$\underset{\text{(in leaves of tree)}}{\text{photosynthesis}} \quad \text{water} + \underset{\text{dioxide}}{\text{carbon}} \xrightarrow{\text{light}} \text{sugars}$$

When sugars have been produced, they are transferred to various parts of the tree via the inner bark, or bast. In other words, movement of liquid up the tree is through the sapwood, and movement of liquid down the tree is through the bast (figure 8.1).

The cambium, which separates the bast from the sapwood, produces new cells for both bast and sapwood when required. The outer bark and the heartwood, which contains fibres, are both dead; that is, there are no living cells present.

Each year, as further rings are added to the thickness of the trunk the amount of dead fibres, or heartwood, increases, and new

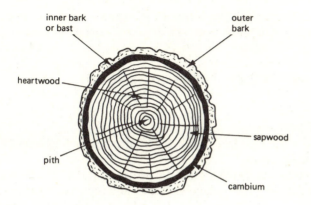

inner bark or bast
outer bark
heartwood
sapwood
pith
cambium

Figure 8.1

Table 8.1

Timber	Density (kg/m^3)
Greenheart (hardwood)	1040
European oak (hardwood)	720
Teak (hardwood)	657
Sycamore (hardwood)	625
English elm (hardwood)	545
Parana pine (softwood)	545
European horse chestnut (hardwood)	497
European spruce (softwood)	465
Western red cedar (softwood)	385

sapwood calls are produced. (As you can see from figure 8.1, the sapwood is that part of the trunk nearest the bark.)

PROPERTIES OF TIMBER

Density and Moisture Content

If we consider timber as bundles of fibres or tubes stuck together (because the cells are long, thin and hollow), it will readily be appreciated that the tubes or cells can hold water, and that the amount of water retained will modify the properties of the timber.

When timber is cut down, it is usually allowed to dry out, or *season*. As the drying proceeds, the density of the timber decreases. The density of a timber is meaningless without a measure of its water content, except as a means of comparing one density with another. Some densities are given in table 8.1, all at 15 per cent moisture content (compare density of water 1000 kg/m^3).

An interesting feature of timber is that it is relatively weak in the wet state and strong when dry. In the drying process, water within the sapwood cells evaporates first. When the moisture content reaches about 27 per cent, the cell walls begin to dry out, and the timber strength continues to increase, reaching a maximum when

the timber is comparatively dry—about 8–12 per cent moisture content, below which embrittlement develops.

Timber used in construction and joinery is rarely completely dry; its moisture content on average is about 10 per cent for interior work, 18 per cent for outdoors—both figures vary depending on the relative humidity of the atmosphere. Thus exterior wooden doors have a lower moisture content in summer than in winter, when they can be difficult to close, owing to excessive water absorption in the wood cells causing swelling.

A *moisture meter*–an instrument containing a calibrated scale, and a probe of two pointed metal prongs—measures moisture contents in timber. The prongs are placed in contact with a specimen of timber and the electrical resistance between them is measured. The resistance is proportional to the amount of moisture present. The scale usually indicates the moisture content directly.

Structural and Mechanical Strength

Timber has obviously established itself as a good, strong, structural material, but, since it has a grain, its strength varies in different planes (see figure 8.2).

Hardwoods are generally denser and stronger than softwoods because the age rings are closer together, that is, the wood cells are

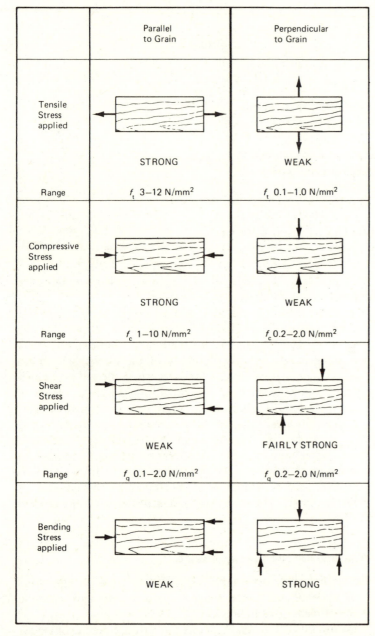

Figure 8.2

packed more tightly. Increases in temperature, and moisture, together with knots tend to reduce the strength of timber.

TIMBER DECAY AND PRESERVATION

Decay

This can occur in three ways: by weathering, by fungus attack and by insect attack.

Exterior timber is subject to weathering; that is, exposure to sunlight can cause fading, bleaching and repeated expansion and contraction. Exposure to rain and other water can cause distortion due to swelling effects. Typical weathering effects are cracking, splitting, surface discolouration and distortion.

Ideal conditions for the growth and development of *fungus* are food (such as timber), moisture, oxygen and a suitable warm temperature (optimum found to be 23°C). This is the sort of environment provided beneath unventilated timber ground-level floors. Three types of fungus are quite common—*mould, dry rot,* and *wet rot* (see table 8.2).

Table 8.2

Type and Cause	Effect	Treatment
Mould: above conditions	Green or black patches on timber surface; harmless	Brush or scrape off
Dry rot: above conditions; once established, it will continue to grow in dry conditions	Growth like cotton wool, on places like undersides of skirting boards. Timber eventually rots, becoming light, porous and brittle	Cut out infected timber, and replace with new treated timber; also treat surrounding timber with preservative
Wet rot: above	Eventually has the	As for dry rot

Type and Cause	Effect	Treatment
conditions plus excess water, found, say, around sinks and tanks	same effect as dry rots, the wood becomes brittle and porous and looses its strength	

Certain insects use timber as food, and can cause serious damage to wooden structures. The commonest ones are given in table 8.3.

Table 8.3

Insect	Damage caused	Treatment
Common furniture beetle or 'woodworm' (*Anobium punctatum*) (60% of all damage in U.K.)	Makes small (2 mm diameter) tunnels or passages in wood, leaving behind little piles of wood dust; eggs are laid on timber surfaces; holes are caused by larvae which hatch out of the eggs	If damage is extensive, the timber must be replaced; otherwise, the holes and the surrounding woodwork are treated with preservative containing an insecticide
Death watch beetle (*Xestobium rufovillosum*)	Extensive damage is caused to hardwoods which have already been softened by decay	Either replace the infected timber or treat extensively with preservative
Powder post beetle (*Lyctus brunneus*)	Causes damage by attacking the sapwood of certain hardwoods; leaves holes and piles of flour-like powder	As for woodworm

Preservation

Most preservatives contain toxic metals and/or polycyclic hydrocarbons, both of which are poisonous to fungi and insects. There are basically three types in use; creosotes, organic solvent types, and water solvent types. The water solvent types are the most effective, because they have the greatest penetration into wood. Many proprietary brands of preservative are available, many of them containing mixtures of chemicals, making them toxic to fungi and insects. Among substances used in preservatives are (apart from creosote)

chlorinated naphthalenes copper naphthenate zinc naphthenate	dissolved in organic solvents, like petrol fractions, before application
copper sulphate zinc chloride potassium chromate	dissolved in water before application

The degree of preservation achieved depends on the amount of penetration into the timber. Creosote is the least penetrating, but is nevertheless a cheap and efficient method of dealing with exterior timber, provided the treatment is renewed every few years. The Post Office treats its telephone poles with creosote by pressure impregnation (see below), and the extra penetration obtained preserves the poles for decades.

Preservatives may be applied in a variety of ways.

(1) *Brushing and spraying*: simple method but not very

effective because penetration is only superficial.

(2) *Dipping or immersing:* the effectiveness of this method depends on the time of immersion of the timber — the longer the immersion, the greater the penetration.

(3) *Pressure impregnation:* the best method available, but the most complex and costly. The timber is placed in a chamber, and vacuum is applied, drawing the air from the pores in the wood. Preservative then floods the chamber and deep penetration or impregnation is achieved. Finally, the vacuum is applied again to remove any excess preservative.

EXERCISES

8.1 Distinguish between a hardwood and a softwood.

8.2 Write an account of fungal attack in woodwork, indicating the principal forms of timber fungus, and the steps that can be taken to deal with it.

8.3 Outline methods used in preserving timber.

8.4 List the advantages of structural timber over structural concrete.

8.5 Give *three* defects occurring in timber. Indicate the effect, cause and remedy in each case.

9 Sound

Vibrations

When something vibrates, that is, moves up and down rapidly about a fixed position, sound is generally produced. If someone thumps the top of the table, you can hear the sound produced because the table top is vibrating to and fro. The vibration of a bee's wings as it flies through the air causes the familiar and ominous buzzing sound. Vibration of the vocal cords produces the sound we know as speech. The crack as a batsman hits the cricket ball is the sound produced by the vibration of the surface layers of the bat. Anything which vibrates tends to cause sound.

The number of vibrations that occur in 1 second is known as the *frequency*. If vibrations of an object occur at less than about 15 times per second, the sound produced cannot be heard by the human ear. Similarly vibrations of above about 16 000 times per second cannot be picked up. All vibrations between 15 and 16 000 produce sound which is audible. This is known as the *range of audibility* for humans. The extreme values vary from ear to ear, but generally the range becomes smaller as a person gets older. Vibrations below 15 are known as *subsonic*, and vibrations above 16 000 are *ultrasonic*.

How Sound is Propagated (or How it Travels)

If we strike a gong we can hear the sound produced. Look at figure 9.1.

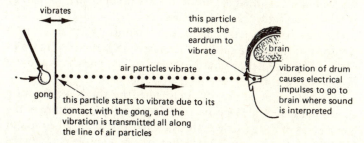

Figure 9.1

Notes and Noises

If one single vibration is produced, the sound heard, which consists of only one frequency, is called a *note*. This happens, say, when a tuning fork is struck. The prongs of the middle C tuning fork, for instance, will move backwards and forwards 256 times a second (256 Hz).

When you thump a table with your fist, different parts of the table surface vibrate at different frequencies. The accumulated effect of these vibrations together is called a *noise*.

Thus when a tin whistle is blown, a single *note* is heard. The *noise* produced at a football match is due to a large number of different frequency vibrations occurring together. Sometimes, however, a number of different vibrations occurring simultaneously produce an over-all pleasant *sound*, as distinct from a noise. A group playing music is an example of this.

A sound source of high frequency produces a high pitched note; a sound source of low frequency produces a low-pitched note. Generally high-frequency sounds are more 'penetrating' than low pitched sounds.

Note: 1 cycle per second is given the name 1 hertz — shortened to 1 Hz.

Waves

Energy can be transferred from one place to another. Thus electricity is sent from the power station to the home or office. Light energy travels 90 000 000 miles from the Sun to the Earth. If, during the transfer of energy through a medium, the particles of the medium are caused to vibrate in a uniform manner, the energy is said to be in the form of a *wave*.

On this basis, sound has a wave nature, because as it is transmitted through a medium, it causes the particles to move to and fro. Figure 9.2 shows the movements of one particle as a sound passes through it.

This behaviour is usually represented by the wave diagram shown in figure 9.3. The wavy line gives a complete picture of the movements of the particle as the wave passes over it. When sound is passing through a medium, all the particles involved perform the same sort of motion.

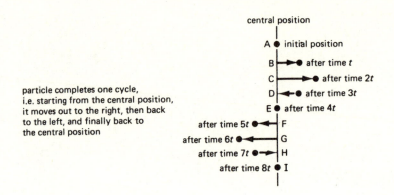

particle completes one cycle, i.e. starting from the central position, it moves out to the right, then back to the left, and finally back to the central position

Figure 9.2

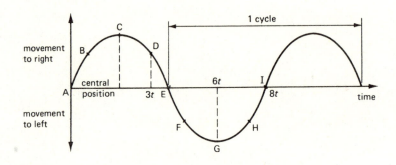

Figure 9.3

There are basically *two* types of wave as follows.

(1) *Longitudinal wave* — such as sound, where the particles vibrate longitudinally, or to and fro in the direction of the wave path, as the wave passes over them.

(2) *Transverse wave* — such as light or radio waves, where the particles vibrate transversely or up and down across the path of the wave, as the wave passes over them (see figure 9.4).

All wave motion, whether longitudinal or transverse, may be represented by the sine curve shown above.

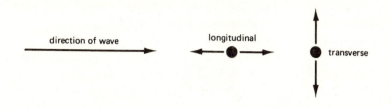

Figure 9.4

Why Sound dies away after a Short Time

Sound is carried forward by the transfer of energy from one particle to another (figure 9.5).

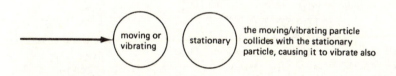

the moving/vibrating particle collides with the stationary particle, causing it to vibrate also

Figure 9.5

The moving or vibrating particle possesses energy of motion, that is, *kinetic energy*. When it strikes another particle, causing movement, kinetic energy is transferred. However, because of the collision, some heat energy is produced; thus

$$\frac{\text{kinetic energy of}}{\text{first particle}} = \frac{\text{kinetic energy of}}{\text{second particle}} + \text{heat}$$

You can see that the second particle has less kinetic energy than the first. This means that the second particle will move more slowly than the first. Eventually, due to successive collisions, the kinetic energy will become zero, and the motion will stop, that is, the sound will die away. It is interesting to note that the original sound energy has all been converted into heat.

SPEED OR VELOCITY OF SOUND

Sound travels at different velocities through different media; see table 9.1.

Table 9.1

Material	Velocity (m/s)
Air	331
Water	1450
Turpentine	1326
Copper	3560
Aluminium	5104

Notice that sound must have a medium through which to travel, since it depends upon molecules and atoms for its propogation; sound cannot travel through a vacuum.

Note from the table that sound generally travels more quickly through denser materials. This is because in dense materials the particles or atoms are packed more closely together so that the transfer of kinetic energy can take place more efficiently. Looking at the table you can see that sound travels through water at approximately four times its velocity through air. Its velocity through aluminium is about sixteen times that through air.

ACOUSTICS

The study of acoustics deals with the design of rooms and halls in buildings to produce the most satisfactory distribution of sound. For example, it is desirable that the sound is not too loud, or too quiet. Good design will make sure that sound is evenly distributed throughout large halls so that the audience can hear what is going on. Sounds should not echo or reverberate around a room or hall, since this will cause confusion.

Obviously this type of study is usually carried out only on halls where large numbers of people congregate to hear something—a speech at a conference, an orchestral concert, a pop group, and so on. It is important that the people at the back can hear, and that the people at the front do not have their ear drums burst by the noise!

Small rooms in private houses do not need such design considerations, but it is as well to be aware of the principles, since they apply equally to small rooms and large halls.

Absorption

When sound strikes a flat surface, it is reflected in a manner similar to light (see figure 9.6). However, the quantity of reflected sound is less because some absorption of sound takes place at the surface. In a closed room sound is absorbed at each reflection until eventually the sound dies away. A sound source, of course, emits millions of sound waves. Only one is shown in the diagram for simplicity. This successive reflection of sound waves is called *reverberation*.

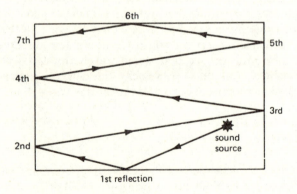

Figure 9.6

Different materials absorb different amounts of sound. No material is available which reflects all sound and absorbs none. The best absorber of sound in a closed room is an open window, since nothing that passes into it comes back. Generally hard, flat surfaces absorb very little sound whereas soft surfaces, like curtains, absorb a high proportion (figure 9.7).

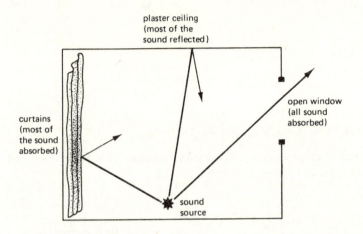

Figure 9.7

Table 9.2

	Material	Absorption Coefficient
These values	Brick	0.02
apply to sound	Plaster	0.03
emitted at 500	Polished wood block	0.06
cycles per sec	Carpet (without underlay)	0.25
(Hz)	Canvas	0.30
	Curtains (medium)	0.30
	Curtains (heavy)	0.70
	Acoustic board	0.70
	Acoustic tiles (on battens)	0.80
	Glass wool (on battens)	0.80

The *absorption coefficient*, which expresses the absorption (per unit area) of a material compared with that of a perfect absorber, has been introduced to compare different materials (see table 9.2).

The amount of absorption of sound by a material also depends on the surface area of the material. Obviously a large curtain will absorb more sound than a small curtain. The relation is expressed by the equation

absorption = absorption coefficient × area (m²)
(in SI units of sound
    absorption)

In some cases, it is more appropriate to quote the total absorption, rather than the absorption per square metre, which is, after all, what the absorption coefficient represents (see table 9.3).

Table 9.3

Object	Absorption (in SI units)
1 person	0.44
1 wood seat	0.20

Reverberation Time

When sound is produced in the open air, it is distributed and dies away in a very short time. On other hand, when sound is produced in a room, it takes longer to die away because reflections from the surfaces in the room cause reverberation and prolong the sound. The *reverberation time* is effectively the time taken for sound to die away in a room or hall. A more precise definition is the time taken for a sound to reduce in intensity by a factor of 1 million, *or the* time taken for the sound level in a room or hall to fall by 60 decibels (see p. 63).

Large churches tend to have long reverberation times. Compare this with a small room containing large areas of curtain and carpeting. Any sound produced is absorbed very quickly, giving a softening or 'dead' effect. It has been found by experience that a half-second reverberation time is most suitable for speech, while a slightly longer time for music, 1 to 2 seconds, gives it quality and richness.

We have already mentioned the deadening effect of too short a reverberation time, Conversely a reverberation time which is too long can cause confusion because new sounds are being produced before the old sounds have died away.

Equations for Calculating Reverberation Time

You will readily appreciate that the time taken for sound to die away in a hall (reverberation time) depends on the volume of the hall and the area of sound-absorbent material present. Several useful formulae have been developed in this connection, and two are given below.

(1) *Stephen's and Bate's formula* is an empirically developed equation designed to give the best (optimum) reverberation time for a given size of hall

$$t = (0.0118 V^{\frac{1}{3}} + 0.107)k$$

where t = optimum reverberation time (s), V = volume of hall (m³) and k = constant (4 for speech, 5 for orchestra, 6 for choir).

(2) *Sabine's formula* gives the actual reverberation time in a given size of hall, given the area and type of absorbent material present

$$t = \frac{0.16 V}{S}$$

where t = actual reverberation time (s), V = volume of hall (m³) and S = absorption (SI units).

When designing a hall for acoustic properties, the aim is to arrange the size of hall and the materials in it so that the actual reverberation time is equal to the optimum reverberation time.

Example in Acoustic Design

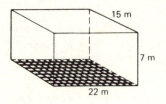

Figure 9.8

Figure 9.8 shows a conference hall designed to accomodate 300 people. The ceiling and walls are plastered and the floor is carpeted. Calculate the area of curtains required so that the optimum and actual reverberation times are equal (absorption coefficients are taken from table 8.2).

Solution   Calculating first the optimum reverberation time

$$t = (0.0118 V^{\frac{1}{3}} + 0.107)k$$
$$= [0.0118(13.22) + 0.107]4$$

$k = 4$ for speech, therefore

$$t = 1.052 \text{ s}$$

For the best design, the actual and optimum reverberation times are to be equal. By Sabine's formula

$$t = \frac{0.16 V}{S}$$

or

$$S = \frac{0.16 V}{t} = \frac{0.16 \times 2310}{1.052}$$

= 352 SI absorption units

This means that 352 units of absorption are necessary to get the best reverberation time. The number of absorption units present already is (area × absorption coefficient)

$$848 \times 0.03 = 25.5 \text{ for plaster}$$
$$330 \times 0.25 = 85 \text{ for carpet}$$
$$300 \times 0.44 = 132 \text{ for people}$$

$$\text{Total} = 242.5$$

Therefore another $352 - 242.5 = 109.5$ absorption units are required, and this is to be provided by curtaining. Therefore

$$\text{area of curtains (heavy) required} = \frac{109.5}{0.7} = 156.5 \text{ m}^2$$

Other Factors in the Acoustic Design of Buildings

The following remarks apply to the distribution of sound into all corners of a hall. The ideal situation is that every member of the audience should hear satisfactorily what is going on. This situation is not always easily achieved, especially in a hall of irregular shape. If this is so, then sometimes projections and awkward looking features such as balconies, pillars and boxes aid sound distribution by providing reflecting surfaces to points where sound would not normally reach.

If amplifying systems are to be used, their positions must be carefully calculated. For instance, loudspeakers should be placed near the speaker, preferably above him (figure 9.9), otherwise members of the audience may hear the same words from two sources with a slight time interval, which is confusing.

An empty space in front of the speaker and a reflecting or sounding board above him help to distribute sound (figure 9.10).

The shape of ceilings can affect the distribution of sound. A curved ceiling can sometimes have a focusing effect so that the

Figure 9.9

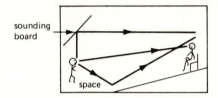

Figure 9.10

sound is concentrated at various points in a room instead of being evenly distributed.

Amplifying systems, although having the beneficial effect of raising the sound level, also unfortunately affect the reverberation time, and this factor must be taken into consideration.

SOUND INSULATION

Unwanted sound or noise is always a nuisance. The greater the intensity or loudness of the sound, the greater the nuisance. If you switch on the radio, it produces a sound, but this is desirable because you want to hear it. However, the motor cyclist who passes you in the street produces unwanted sound and is a nuisance. If you live in a flat and the man above is always practising on his trumpet for the local works band, you are annoyed by the unwanted intrusion of noise in your life. Noise can be a pollutant just like litter or smoke. It is the purpose of sound insulation to cut out or reduce unwanted sounds or noise to a minimum.

It is very strange that there are *two* distinct contradictory types of insulating material, that is, material which will cut out or minimise sound or noise. One type consists of very heavy dense materials, such as brick, concrete and stone, while the other is composed of very light, porous materials such as fibreglass, and expanded plastics foam. Both kinds of material perform ultimately the same purpose, though by different mechanisms.

Dense Materials (figure 9.11)

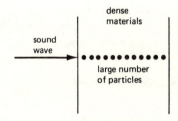

Figure 9.11

The sound is transmitted through the wall via the particles of the wall material. If the wall is dense, there are so many particles packed together that the sound dies away in the wall by successive collisions.

Diffuse Materials (figure 9.12)

Diffuse materials are full of long passages containing air. Collisions of air particles in the passages transmit the sound, which dies away in the passage by successive reflections off the walls of the passage.

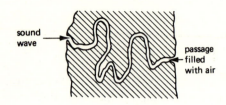

Figure 9.12

Remember that sound is transmitted through materials by collisions and vibrations of the respective particles. Each time a collision occurs, some heat is generated and the new vibrating particle has less energy. Eventually the particle motion (that is, the sound) dies away.

In the dense materials, the motion dies away in a fairly short length because the particles are packed tightly together.

In the diffuse materials, the particles are packed less tightly but the over-all effect is the same because the passages are very long.

Basic Principles of Sound Insulation

If possible keep the source of sound as far away as possible from a building, because sound obeys an *inverse square law* relationship, that is, the intensity of sound arriving at a certain point is inversely proportional to the square of the distance from the sound source to the point (see chapter 11). For example, if a new building is to be built on a busy road, it should be located as far back from the road as possible. If this is not possible, a wall should be constructed or bushes or trees planted between the road and building to obstruct the sound (figure 9.13).

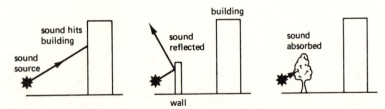

Figure 9.13

Make the exterior walls as dense as possible to prevent sound entering the building from outside, and double glaze windows, since these are more efficient sound insulators than single glazed windows.

On the whole, gaps and cracks in door and window openings allow considerable sound through them, and should be treated to render the building sound-proof.

Once everything has been done to prevent sound entering the building from outside, attention must be paid to the problem within the building, since sound also travels through partition walls, ceilings and floors into adjoining rooms.

It is desirable to place quiet rooms, such as offices, well away from noisy rooms, such as factory floors where machinery may be in use. Heavy machinery is cumbersome and is usually located on the ground floor because of its weight. This is also a desirable position from the point of view of sound insulation since vibration would be greater on higher storeys. Even so, most heavy vibrating machinery is mounted on rubber pads to further prevent vibration nuisance.

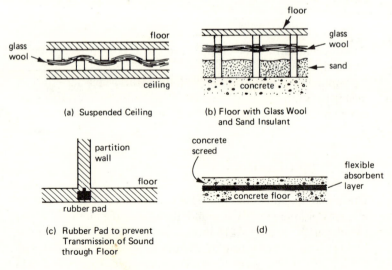

(a) Suspended Ceiling

(b) Floor with Glass Wool and Sand Insulant

(c) Rubber Pad to prevent Transmission of Sound through Floor

(d)

Figure 9.14

Various constructions are available for insulating sound in a building. Apart from dense partitions, those shown in figure 9.14 are all in use.

SOUND OR NOISE LEVEL

The transmission of sound through the air causes vibrations of air particles. Particles close to the ear drum hammer against it, causing it to vibrate as well, and the brain interprets the vibrations so that the meaning of the sound is understood.

The ear particles vibrating against the ear drum exert a pressure on it. If the loudness or intensity of the sound is great, great pressure is exerted on the ear drum; that is, the pressure experienced by the ear drum is proportional to the loudness or sound level being experienced.

A pressure of 0.000 02 N/m² is just detectable by the human ear and is referred to as the *threshold of hearing*. This pressure may also be referred to in terms of energy, and is 0.000 000 000 001 W/m², or 10^{-12} W/m². These are called respectively the *threshold pressure* and *threshold energy*.

The other extreme is where the level of sound is very high, and it is called the *threshold of pain*. Beyond this level the ear drum would be permanently damaged. Values here are 200 N/m² pressure and 1 W/m² energy intensity.

It will readily be appreciated that both the pressure and energy ranges are extremely wide.

	threshold of hearing		*threshold of pain*
pressure (N/m²)	0.00002	→	200
energy (W/m²)	0.000 000 000 001	→	1

These ranges can be simplified into a simple scale, the *decibel* (dB) scale

0 dB → 120 dB
threshold of hearing threshold of pain

The formulae used to convert pressure and energy intensities into *sound pressure levels* (in decibels) are as follows.

$$\text{Sound pressure level (dB)} = 10\log_{10}\left(\frac{I_1}{I_0}\right)$$

where I_0 = energy intensity at threshold of hearing = 10^{-12} W/m² and I_1 = energy intensity of the sound in question.

$$\text{Sound pressure level (dB)} = 20\log_{10}\left(\frac{P_1}{P_0}\right)$$

where P_0 = pressure intensity at threshold of hearing = 0.00002 N/m² and P_1 = pressure intensity of the sound in question.

Sound Levels of Some Common Noise Sources

Table 9.4

Noise Source	SPL (dB)	
Dropping a pin	15	
General background noise	30	
Heavy falling rain	50	These values are
Sitting in a moving car	50	approximate, but
Conversation	70	they serve to give
Shouting	80	some idea of
Discotheque	90	relative noise levels
Motor bike	95	
Pneumatic drill	110	

Calculation of Sound Pressure Levels

Provided that the energy intensity or pressure intensity of a sound is known, it is a simple matter to find the sound pressure level in decibels, simply by substitution into the appropriate formula.

Example 9.1

A noise has an energy intensity of 0.5 W/m². Calculate the sound pressure level in decibels.

Solution

$$SPL = 10 \log_{10}\left(\frac{I_1}{I_0}\right)$$

$$= 10 \log\left(\frac{0.5}{10^{-12}}\right)$$

$$= 10(\log 0.5 - \log 10^{-12})$$

$$= 10(\bar{1}.699 - \bar{1}2)$$

$$= 116.9 \text{ dB}$$

Adding Sound Levels Together

If a motor bike passes you in the street producing a sound level of, say, 80 dB, and simultaneously another motor bike passes in the opposite direction, your ear is subject to pressure from two sources at the same time. The resulting sound pressure level is not the sum of the two sounds, 160 dB, or your eardrums would burst. In fact, it is a matter of common experience that the noise produced by the second motor bike has very little effect. The sound pressure levels are not simply added together, because of the logarithmic relationships involved.

If you are standing by a pneumatic drill, of, say, 100 dB noise level, and a car, of 60 dB, passes you, the noise effect of the car is barely noticeable. The resulting noise level is more like 101 than 160 dB!

The method of approaching the problem is to calculate the pressure intensity or energy intensity of each of the sounds occurring simultaneously, and sum them to get a total pressure or energy. Substitution of this total into the equations will give the combined sound pressure level.

Example 9.2

Calculate the total sound pressure level due to two pneumatic drills operating together, each of 110 dB sound level.

Solution

$$SPL = 10 \log_{10}\left(\frac{I_1}{I_0}\right)$$

$$I_0 = 10^{-12} \text{ W/m}^2$$

$$110 = 10 \log\left(\frac{I_1}{10^{-12}}\right)$$

so

$$11 = \log I_1 - \log 10^{-12}$$

or

$$11 = \log I_1 + 12$$

$$\log I_1 = -1 \text{ or } \bar{1}.0000$$

$$I = 0.1 \text{ W/m}^2$$

Therefore the total energy intensity reaching the ear due to the two drills is $0.1 + 0.1$, that is, 0.2 W/m². It is now a simple matter to recalculate the resultant sound pressure.

$$SPL = 10 \log\left(\frac{I_1}{I_0}\right)$$

$$= 10 \log\left(\frac{0.2}{10^{-12}}\right)$$

$$= 113.0 \text{ dB}$$

Note that the second pneumatic drill increases the sound pressure level by only 3 dB.

A third pneumatic drill of 110 decibles would cause a further addition of 0.1 W/m² to the energy intensity, giving a total $0.1 + 0.1 + 0.1 = 0.3$ W/m². The sound level then becomes

$$10 \log\left(\frac{0.3}{10^{-12}}\right) = 10\ (\log 0.3 - \log 10^{-12})$$

$$= 10(\overline{1}.4771 + 12)$$

$$= 10(11.4771)$$

$$= 114.8 \text{ dB}$$

Note that the third pneumatic drill increases the sound level by only 1.8 dB.

ULTRASONIC AND SUBSONIC SOUND

Objects that vibrate or move to and fro at less than fifteen times per second, or greater than 16 000 times per second are outside the range of human audibility and therefore cannot be heard. Those below fifteen are *subsonic*, and those above 16 000 are *ultrasonic*.

High-frequency or ultrasonic vibrations have many applications, although outside the hearing range. For example, they have been used for killing bacteria, and for welding plastics together, as shown figure 9.15.

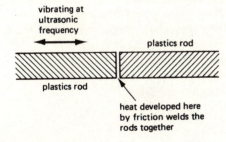

Figure 9.15

Ultrasonic vibrations may be produced by a phenomenon known as the *piezoelectric effect*. It has been found that when an alternating electric current is applied to the faces of some crystals, notably quartz, the faces of the crystal vibrate at very high frequency. Frequencies of the order of 1 000 000 cycles per second (Hz) are known.

Subsonic vibrations have recently been applied to drilling operations, particularly for oil.

EXERCISES

9.1 What is meant by the following?
(a) Sabine's fomula
(b) ultrasonic sound
(c) range of human audibility
(d) sound
(e) acoustics.

9.2 A hall of size 20 m × 60 m and height 10 m is to be built, basically for the performance of orchestral concerts. Calculate the optimum reverberation time for this hall. Calculate also the size of the audience which can be accommodated if the floors are carpeted, and the ceiling and walls plastered, assuming that the actual reverberation time is equal to the optimum reverberation time. (Absorption coefficients: plaster 0.03, carpet 0.5, audience (per person) 0.44 SI units of absorption; k for orchestra = 5).

9.3 A lecture room is 7 m × 7 m × 3 m high. The absorption coefficients at 500 Hz frequency are as follows

walls, hard plaster on brick 0.02
floor, cork tiles 0.05
windows, 12 m² in walls 0.10
ceiling, plasterboard 0.10

There are also 30 padded seats in the room at 0.2 SI units of absorption each. Determine the reverberation time of the lecture room.

9.4 Why is it that dense materials and light porous materials are both very efficient absorbers of sound?

9.5 It is desired to reduce the reverberation period in a lecture theatre from 2.7 s to 1.4 s. Calculate the area of absorptive tiles required to do this, given that the absorption coefficient of the tiles is 0.6. See figure 9.16.

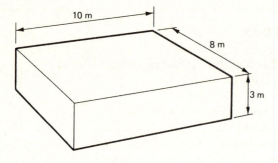

Figure 9.16

9.6   Two cars, one of sound level 80 dB and the other 72 dB pass each other. Calculate the resulting sound pressure level.

9.7   Two cement mixers each have noise levels of 85 dB. Calculate the resulting sound level when both are being used simultaneously.

9.8   Three noise sources produce 78 dB, 84 dB and 70 dB. Calculate the resulting sound level when the noises occur together.

9.9   Explain why it is, when a single noise (A) is produced at a certain sound level, that on producing two such noises (2 × A) the human ear notices only a small increase in the noise level and not twice as much.

   If you had a machine running and generating noise in a building, at what frequency would you run it to produce the minimum disturbance, and where would be the best place to site it? Give your reasons in each instance.

9.10   (a) Distinguish between *echo* and *reverberation*.

   (b) Define the reverberation time of a hall and state briefly how good reception (i) of speech, (ii) of music, depends on the magnitude of this quantity.

   (c) Discuss and explain how the reverberation time of a hall is affected by (i) its size, (ii) its shape, (iii) the nature of its wall surfaces, and (iv) the size of an audience.

9.11   (a) Describe and explain, with examples, the principles to be followed when insulating a building against structure-borne and airborne sound. At least three examples of methods used for each type of sound should be given. Sketches may be used to illustrate answers.

   (b) Describe briefly *four* methods of improving the acoustics of a large noisy workshop or of a lecture theatre.

9.12   A small engineering factory consisting of workshops and offices is to be built in one block of buildings, on a busy trunk road. Indicate and discuss the steps that should be taken to insulate the offices from sound disturbances from traffic outside and workshop noise inside.

9.13   (a) Define fully with respect to sound
   (i) frequency
   (ii) absorption coefficient
   (iii) Stephen's and Bate's formula.
Include, where appropriate, response limits of the human ear.
   (b) With the aid of sketches explain how to
   (i) improve the hearing of wanted sounds in a building
   (ii) prevent all types of unwanted sound entering and spreading through a building.

10   Electricity

CREATION OF ELECTRICITY

As we have said earlier (see chapter 2) an atom consists of a
nucleus, with negatively charged electrons circulating around, or
orbiting, it. If we can disengage electrons and cause them to flow or
move along a conductor in a certain direction, we have created a
current of electricity. Even a small current involves millions of
electrons. The electrons proceed down a wire by a series of jumps
from atom to atom (figure 10.1).

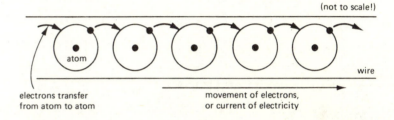

Figure 10.1

Static electricity is also composed of electrons, but in this case
they are moving randomly, if at all, and not in a uniform direction,
as is the case with current electricity.

The most important methods of producing electricity are

(1) frictional methods (for static electricity)
(2) thermoelectric effect
(3) photoelectric effect
(4) generator or dynamo
(5) electrolytic cell
(6) piezoelectric effect.

Frictional Methods

When two surfaces are rubbed together, electrons are freed from
atoms, and static electricity is produced. For instance, if a plastics
comb is drawn through hair the static electricity produced may be

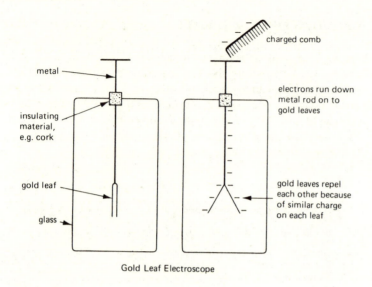

metal

insulating
material,
e.g. cork

gold leaf

glass

charged comb

electrons run down
metal rod on to
gold leaves

gold leaves repel
each other because
of similar charge
on each leaf

Gold Leaf Electroscope

Figure 10.2

detected by an instrument called a gold leaf electroscope (figure 10.2). Producing electricity this way has no commercial feasibility.

Thermoelectric Effect

It was discovered by Seebeck in 1826 that if two wires of dissimilar metal are joined together at two junctions, and one junction is made hotter than the other, a current of electricity passes through the wires. This phenomenon is known as the *Seebeck effect* (figure10.3).

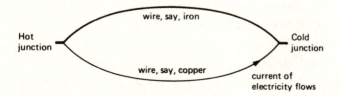

Hot
junction

wire, say, iron

wire, say, copper

Cold
junction

current of
electricity flows

Figure 10.3

The amount of current produced depends on the metals used, and the temperatures of the hot and cold junctions. In general, the greater the temperature difference, the greater is the current flowing. As with the frictional method, this method is not used to produce electricity on a large scale.

The reverse process, discovered independently by Peltier, and known as the *Peltier effect*, is that if a current is passed through a circuit containing two dissimilar metals, one junction becomes hot and the other cold.

The most important application of these principles is that of a thermometer, since the current involved is proportional to the junction temperatures. Such a thermometer, called a thermoelectric thermometer, is sometimes used to measure furnace temperatures in industry.

Photoelectric Effect

When light falls on to the surface of some materials, notably selenium, electrons are released from some of the atoms, and can be caused to flow, producing an electric current (figure 10.4). This is the photoelectric effect (Greek *photos*, light). The electrical energy produced can be considerable, but the principle is not used to produce large scale amounts of electricity.

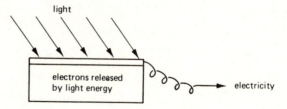

light

electrons released
by light energy

electricity

Figure 10.4

Generator or Dynamo

This is an extremely important method of producing electricity, both alternating current (a.c.) and direct current (d.c.) Nearly all large scale electricity production in the world is based on this method. Most power stations, whatever their source of energy–gas, coal, oil, water or nuclear–generate electricity in this way.

The basic principle is that *whenever a conductor moves in a magnetic field a current flows in the conductor* (figure 10.5).

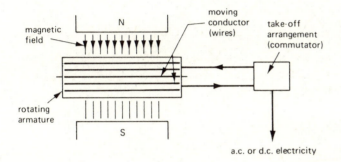

Figure 10.5

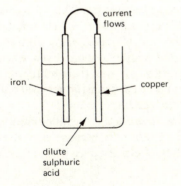

Figure 10.6

Electrolytic Cell

When any two dissimilar metals are placed in an, electrolyte solution (a solution which conducts electricity), and connected by a wire, a current of electricity flows round the circuit (figure 10.6). (See also chapter 2.)

The potentiality of the *cell* produced, or its electromotive force (e.m.f.), may be calculated by reference to the electrochemical series, which arranges metallic elements in the order of their respective *standard electrode potentials*, table 10.1.

Table 10.1

Element	Standard Electrode Potential (V)
Potassium	−2.92
Sodium	−2.71
Aluminium	−1.66
Zinc	−0.76
Iron	−0.44
Hydrogen	0.0
Copper	+0.34
Gold	+1.50

The e.m.f. of any cell is the difference between the two standard electrode potentials of the two metals concerned. Thus the e.m.f. of the cell in figure 10.6 is $+0.34 - (-0.44) = 0.78$ volts. Obviously, from the figures given above, the best cell would be a potassium–gold cell, but this is impracticable because potassium reacts violently with water, and gold is very expensive. As you probably know, other cell defects such as polarisation and local action also limit the number of feasible and useful cells.

Nevertheless, this source of electricity is very useful for small local supplies — the lead – acid accumulator for cars, and the dry cell for torches.

Piezoelectric Effect

It was mentioned in chapter 9 on sound that if an alternating current is applied across the opposite faces of a quartz crystal, the crystal vibrates at high frequency. This piezoelectric effect can also be applied in reverse to obtain electricity. That is, by causing the crystal to vibrate, an alternating current may be produced. Again, this principle is not used for large-scale production of electricity.

ALTERNATING CURRENT

Electricity is a difficult energy form to handle because it cannot be stored. It has to be produced, transmitted and used in a short space of time. Electricity is produced at large power stations, almost invariably by means of the generator technique mentioned above. Although direct current and alternating current can be produced equally easily, alternating current production is carried out because it is generally easier to transmit over long distances in the most efficient way.

Alternating current is so called because the electrons build up in one direction, then die away, and build up in the opposite direction—all in a short space of time. This can be represented as shown in figure 10.7. A typical frequency time is 50 cycles per second (Hz). That is, the current reverses direction 50 times each second.

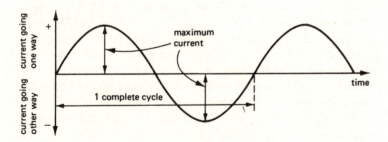

Figure 10.7

Figure 10.7 illustrates what is known as *single-phase alternating current*. Sometimes, in order to obtain a smoother and more

effective supply of electricity, *three-phase a.c.* is used. Here three a.c. supplies are combined together, either all taken from the same generator or from three different generators, The three alternating currents are out of phase with each other by the same amount, as seen in figure 10.8

Figure 10.8

When an electric current passes through a wire, heat is produced, that is, some electrical energy is wasted through conversion into heat. Alternating current is used for long-distance transmission through wires, sometimes hundreds of miles, from generator to electrical appliance. This is because of a piece of equipment called a *transformer*, which can take an electrical supply and, without much loss of energy, convert it to a very high voltage low current supply system. Even when this is carried hundreds of miles through wires, there is little energy loss, because of the low current involved. Transformers will only operate on a.c. systems, *not* on d.c. systems.

What this means, in essence, is that alternating current is used in conjunction with transformers, because this is the most convenient way of transmitting electricity over long distances.

THE TRANSFORMER

Most transformers consist of two insulated coils, or windings, of wire wound round an iron ring or core. Their purpose is to raise or lower the voltage of an alternating current.

A *step-up* transformer raises the voltage; a *step-down* transformer lowers the voltage.

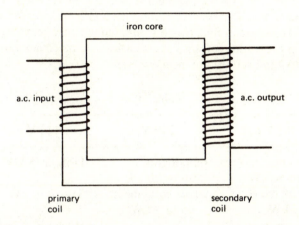

Figure 10.9

When an alternating current is passed into the primary winding (figure 10.9), another alternating current is induced in the secondary winding by the principles of electromagnetic induction. The voltage induced in the secondary winding depends only on the primary voltage and the number of turns of wire in each of the primary and secondary windings, thus

$$\frac{\text{voltage in secondary}}{\text{voltage in primary}} = \frac{\text{number of turns in secondary}}{\text{number of turns in primary}}$$

Thus in a step-up transformer, where the output voltage is greater than the input voltage, the secondary has more turns than the primary. This device is particularly useful for raising the voltage from the generator at the power station prior to feeding into the National Grid for transmission. A step-down transformer, on the other hand, is a useful device for reducing a.c. voltage. In this type, the primary has more turns than the secondary. This type of transformer is also used in the National Grid system.

A transformer has the disadvantage that it is not perfectly efficient, and some electricity is lost as heat. The efficiency of a transformer is given by

$$\text{Transformer efficiency} = \frac{\text{output power}}{\text{input power}}$$
$$= \frac{\text{current} \times \text{voltage (secondary)}}{\text{current} \times \text{voltage (primary)}}$$

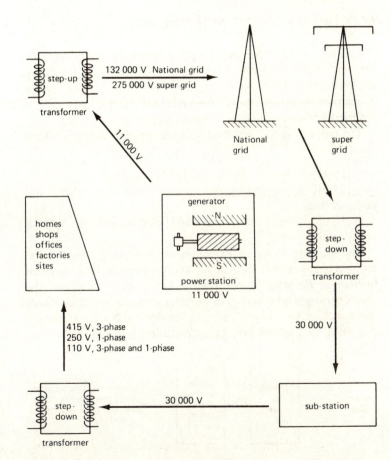

Figure 10.10

THE NATIONAL GRID

Electricity is generated at power stations, and in the United Kingdom there are many of these dotted about the country. The electricity produced is distributed to all parts of the British Isles through a vast network of wires carried on pylons and some underground cables. This is the National Grid System, which is best illustrated by figure 10.10.

ELECTRICITY ON THE BUILDING SITE

The introduction of electricity on to a building site is potentially very dangerous. Since this electricity is to be used only in operating the machines and tools for the construction in hand, it is obviously a temporary arrangement. Nevertheless, careful planning is required because electricity, if mishandled, can be a killer.

The types of supply required on a site where building is taking place are

(1) 415 V 3-phase 50 Hz a.c. for heavy plant
(2) 240 V single-phase 50 Hz a.c. for site offices and floodlighting
(3) 110 V single- and 3-phase 50 Hz a.c. for local site lighting and portable tools.

The *Building Research Establishment* has carried out a large amount of research in this subject and their conclusions are given in *Building Research Station Digest No.87*. They recommend the use of strong steel cabinets or cubicles to house essential equipment such as the supply authority's cable, intake fuses, meters, transformers, and out-going circuit-control gear (figure 10.11).

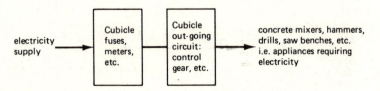

Figure 10.11

If possible, most electrical equipment is nowadays made to operate at 110 V, because this reduced voltages is less dangerous on site where there is much exposure to wet or damp conditions. A whole range of 110 V equipment is now available. Some heavy equipment, however, must operate at the higher voltage (415 V) to avoid excessive currents, which would require very heavy cables.

The power ratings of some site electrical equipment are given in table 10.2 (all ratings are in kilowatts).

Table 10.2

415 V a.c.	240 V a.c.	110 V a.c.
Tower crane motors, up to 60 kW	Floodlights, up to 20 kW	Drill, 0.75 kW
Concrete mixer, 1–15 kW	Office equipment, up to 3 kW	Saw, 0.30 kW
Pump, up to 30 kW		Hammer, 0.75 kW
Concrete saw, 2 kW		

Cables

Some recommendations have been made with regard to wires or cables on site. It is recommended initially that all cables should be 25 m long, with a plug at one end and a socket at the other. This means that if the cable snaps, then all that is required is for that particular cable length to be replaced. This avoids any wire joining, which is always unsatisfactory. All work should be carried out by qualified electricians, and to recognised safety standards. Also in this connection, wiring should comply with British Standards. Cables should be positioned so that they cannot be run over by vehicles, or become worn by dragging on the ground.

EXERCISES

10.1 (a) Draw a diagram of a transformer and explain how it works.

(b) A transformer has 25 turns on the primary and 200 turns on the secondary. The voltage and current in the primary are 10 V and 5 A respectively. If the transformer is 90 per cent efficient, calculate the current in the secondary.

10.2 Write a concise report about general safety conditions relevant to a building site.

10.3(a) Show by means of a fully detailed flow diagram, how electricity is distributed by the National Grid. Explain with a simple diagram, the principle of an alternating-current generator.
   (b) State *five* important points to bear in mind when installing an electricity supply on a building site.

10.4(a) With the aid of a sketch explain how mechanical energy can be converted into alternating-current electricity.
   (b) Sketch and explain with notes the system of distribution of electrical power to a moderate-sized town.

10.5(a). A step-down transformer in a substation has 2200 turns on the primary and 50 turns on the secondary . If the secondary current is 15 A , calculate the current in the primary coil if the transformer's efficiency is 85 per cent.
   (b) Give reasons why the National Grid uses a high-voltage low-current system for transport of electricity.

10.6 Describe how electrical power should be distributed to a medium-sized building site, mentioning the important points to bear in mind, and the equipment required.

11 Light

PHOTOMETRY

When we build a house, factory, shop or block of offices, it is most important that the interiors be well-lit or illuminated. There is no tax on window area nowadays, as there has been in the past, and daylight is free for everyone.

A room may be illuminated entirely by daylight at daytime and artificial light at night. Many modern shops and other large buildings rely partly on daylight and partly on artificial light for their illumination. It is important to have adequate lighting, and the purpose of photometry, as its name suggests, is to measure light or illumination. Photometry is a precise science, and from it we can design correct illumination levels in buildings.

Definitions (figure 11.1)

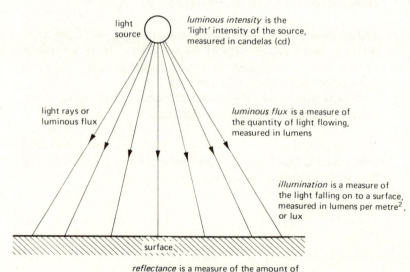

light source

luminous intensity is the 'light' intensity of the source, measured in candelas (cd)

light rays or luminous flux

luminous flux is a measure of the quantity of light flowing, measured in lumens

illumination is a measure of the light falling on to a surface, measured in lumens per metre^2, or lux

surface

reflectance is a measure of the amount of light reflecting from an illuminated surface: expressed as a percentage of the total incident light

Figure 11.1

Lumen (figure 11.2)

The light or luminous flux emitted from a light source of 1 candela in unit solid angle is called a *lumen*.

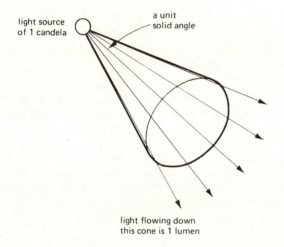

Figure 11.2

Lux (figure 11.3)

When a luminous flux of 1 lumen falls on to 1 m² area of surface, the illumination produced is 1 *lux*. (1 lux = 1 lumen per square metre.)

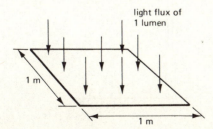

Figure 11.3

Candela

At one time a candle was used as a standard source of light, but since candle flames vary in size and intensity, its accuracy was questionable. Hence there was a need to find a better standard.

When bodies are heated, they emit radiation. At high temperatures, much of the radiation is in the form of light: for instance, a red hot poker, a light filament. One square centimetre surface of a body at 2046 K is defined as having a luminous intensity of 60 *candelas* (2046 K is the melting point of platinum)—see figure 11.4. So a candela is defined as a unit of luminous intensity such that the luminous intensity of a full radiator at 2046 K is 60 candelas per square centimetre. In other words, the light emitted from a white hot body such as platinum is more constant than that given out by a candle flame.

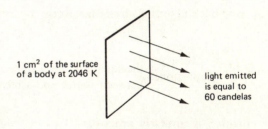

Figure 11.4

Inverse Square Law

In a dimly lit street at night, it is difficult to see, and the brightest place is underneath a street lamp. As you walk away the light fades. The illumination or brightness is obviously related to the distance from the light source (figure 11.5).

According to the inverse square law, *the illumination of a surface due to a light source is inversely proportional to the square of the distance between the source and the surface.*

For example, in figure 11.5, let the illumination on the first screen be 100 lux. The second screen is 2 m away, so the illumination on it

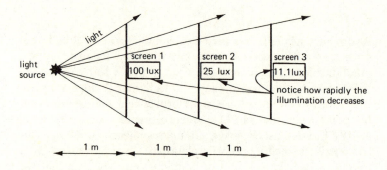

Figure 11.5

is $1/2^2$, one-quarter that of the first, or 25 lux. The third screen, 3 m from the light scource, has an illumination of $100/3^2$, that is, $100/9 = 11.1$ lux.

We can develop a useful formula from the inverse square law as follows.

Illumination on a screen	is inversely proportional to	the square of the distance between source and screen

that is, illumination is inversely proportional to the square of distance, or

$$\text{illumination } E \propto \frac{1}{(\text{distance } d)^2}$$

so

$$E \propto \frac{1}{d^2}$$

or

$$E = \text{constant} \times \frac{1}{d^2}$$

It is found that the constant in this equation is equal to the luminous intensity of the source, designated I. Therefore

$$E = I \times \frac{1}{d^2}$$

or

$$E = \frac{I}{d^2}$$

where E = illumination of surface (lux), I = luminous intensity of source (candela) and d = distance between source and screen (m).

Example 11.1

Calculate the illumination on the road at a point A in figure 11.6, directly beneath the lamp.

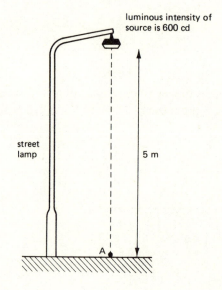

Figure 11.6

Solution

$$E = \frac{I}{d^2} = \frac{600}{5^2} = 24 \text{ lux}$$

Therefore illumination at A is 24 lux.

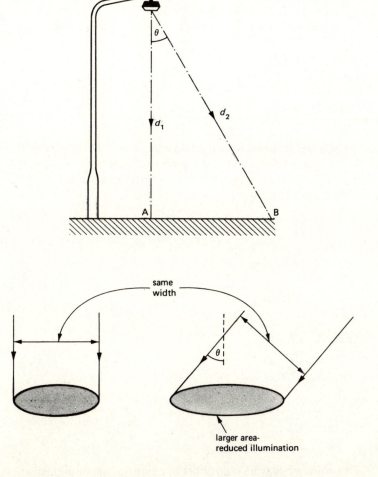

Figure 11.7

Lambert's Cosine Law

The equation used above is quite suitable for use when the light falls *perpendicularly* on to the surface, as it did in the above example. If, however, the light falls obliquely on to the surface, the equation must be modified. In figure 11.7, the light falls normally on to A, but obliquely on to B. The illumination at B is less that at A because

(1) distance d_2 is greater than distance d_1
(2) light falling obliquely on to a surface covers a larger area with a correspondingly reduced illumination.

The modified equation for light falling obliquely is called Lambert's cosine law

$$E = \frac{1}{d^2} \cos \theta$$

θ is the angle indicated in figures 11.7 and 11.8; note that θ is the angle the rays make with the perpendicular to the surface.

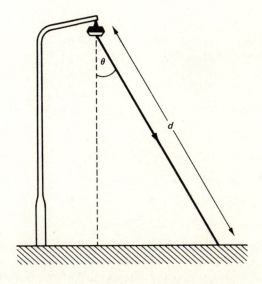

Figure 11.8

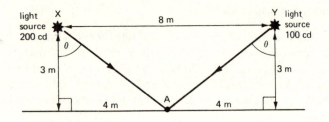

Figure 11.9

Example 11.2

Calculate the illumination at A in figure 11.9 due to the light sources.
Solution

Illumination at A, due to source $X = \dfrac{I}{d^2} \cos \theta$

$$= \frac{200}{5^2} \times \frac{3}{5}$$

Illumination at A, due to source $Y = \dfrac{100}{5^2} \times \dfrac{3}{5}$

Total illumination at A $= \dfrac{200}{5^2} \times \dfrac{3}{5} + \dfrac{100}{5^2} \times \dfrac{3}{5}$

$$= \frac{24}{5} + \frac{12}{5}$$

$$= 7.2 \text{ lux}$$

Photometers

The inverse square law and Lambert's cosine law are both useful in designing artificial lighting in buildings. Photometers are instruments used for measuring and comparing the luminous intensities of sources of light. Tables are available for converting candela ratings into wattage ratings, which are more usual for lamps. Many

different types of photometer are available, including the Bunsen photometer, Joly's photometer, the flicker photometer and the Lummer–Brodhun photometer. Only a few simple ones will be considered here (see figures 11.11 and 11.12).

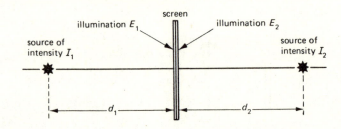

Figure 11.10

Photometers operate on the principle shown in figure 11.10. The screen is two sided, and the *two sources of light are moved towards or away from the screen until the illuminations E_1 and E_2 are the same.*
Now

$$E_1 = \frac{I_1}{d_1{}^2}$$

and

$$E_2 = \frac{I_2}{d_2{}^2}$$

Since $E_1 = E_2$

$$\frac{I_1}{d_1{}^2} = \frac{I_2}{d_2{}^2}$$

or

$$\frac{I_1}{I_2} = \frac{d_1{}^2}{d_2{}^2}$$

This equation may be used either to compare luminous intensities of two sources, or to measure the luminous intensity of one source.

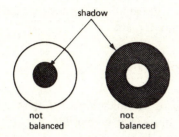

Figure 11.11

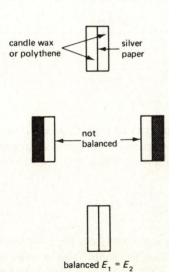

Figure 11.12

Bunsen Photometer (figure 11.11)

The screen is simply a filter paper with a central grease spot.

Joly's Photometer (figure 11.12)

SOURCES OF LIGHT

There are basically three types of light source
  (1) incandescent
  (2) gaseous discharge
  (3) fluorescent.

Incandescent Lamp (figure 11.13)

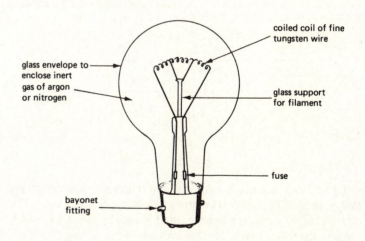

Figure 11.13

When an electric current is passed through the tungsten filament it becomes white hot (incandescent) and so emits light. If this were done in the atmosphere the filament would rapidly oxidise or burn away, hence the purpose of the glass envelope and inert gas is to protect it.

The filament when in operation is at a temperature of about 2750°C, which in practical terms means that most of the radiation emitted is heat, and that only about 10 per cent of the electrical energy used is converted to light. The efficiency of the lamp is therefore low.

Nevertheless, due to its ease of construction, cheapness, and good colour rendering, the lamp is very popular. (Colour rendering is the term used to compare how the type of light emitted from a lamp compares with daylight).

Gaseous Discharge Lamp

Sometimes when an electron strikes an atom, the atom ionises, and at the same time releases energy in the form of visible light. In the gaseous discharge lamp the atoms are trapped in a glass tube (figure 11.14) and electrons are sent along the tube from terminal points, called electrodes, fitted at the ends of the tube.

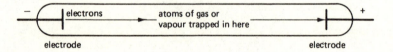

Figure 11.14

These lamps are much more efficient than incandescent lamps, that is, their light output is greater per unit of electricity used. The disadvantage is that the light emitted is usually coloured, which renders them unsuitable for domestic and interior use.

The vapours most commonly used for filling the lamps are sodium (yellow light) and mercury (bluish light).

This type of lamp is particularly suitable for street and highway lighting, where good illumination is necessary but colour is not so important.

Sodium and mercury lamps both contain a little argon or neon gas as a starter, since neither sodium nor mercury is a vapour at normal temperatures. After the lamp has been on for a little while, and has warmed up, the sodium and mercury vaporise, and normal functioning then takes place.

Fluorescent Lamp

As we said above, when an electron strikes an atom and causes ionisation, energy is emitted. Some of this emitted energy is ultraviolet radiation, which is invisible, but if it strikes certain fluorescent powders, light is given out. In the fluorescent lamp (figure 11.15) mercury vapour at low pressure provides the atoms which become ionised, and they are placed in a long glass tube with electrodes at each end. The inside surface of the glass tube is coated with the fluorescent material.

Various fluorescent powders are available, emitting different colours. If these powders are mixed, a light of good colour rendering may be obtained, that is, the light emitted is very similar to daylight. In fact, fluorescent lamps or tubes may be obtained in a

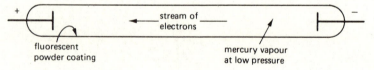

Figure 11.15

range of different colour renderings, including white, warm white (a little more red in it), northlight (more blue in it) and natural light. These lamps are much more efficient than incandescent lamps

LIGHTING DESIGN

A room in a building is generally designed to have a good level of illumination throughout the room. This may be obtained using daylight through windows, artificial light from lamps, or a mixture of the two. If the light is too dim you may get eye-strain; if too strong you may get a headache because of glare.

Photographic dark-rooms, of course, do not need windows. On the other hand a greenhouse, which is all windows, lets in plenty of light but at the same time it can get very hot because it also allows heat in. Light entering a room through a window is often obstructed by objects outside, such as trees and buildings. However, a room made entirely with windows or lights in the ceiling or roof would be unobstructed but none the less, unsatisfactory, because we also like to use windows to look outside and see what is going on.

We design lighting purely for our own comfort and convenience.

Daylighting

Apart from open doorways, all daylight enters rooms through glass windows. Obviously the bigger the windows, the better the illumination in the room. However, since rooms also have to be heated, and in cold weather a large amount of heat is lost from buildings through windows, a compromise has to be reached. A window must allow in adequate light, but must not be so large that an excessive amount of heat is lost through it.

Another point to note is that as light passes through glass, a little of it is absorbed (up to 10 per cent) so that effectively less light comes through a closed window than an open one. This is known as the *transmission loss* of light.

The simplest method of measuring the illumination in a room is to use a *lightmeter*, which is simply a photoelectric cell (see p. 68) that generates electricity when light falls on to it. The current produced operates a pointer which moves over a scale recording the illumination in lux units. It is virtually impossible to quote recommended illumination values in buildings lighted by daylight, because obviously the illumination inside depends on the illumination outside, and in the United Kingdom the weather, and brightness, vary tremendously from day to day. Compare the illumination in a kitchen on a sunny summer day with that on a gloomy winter day.

To overcome this, the illumination in a room is expressed as a percentage of the illumination outside. This percentage, called the *daylight factor*, is given by

$$\text{daylight factor} = \frac{\text{illumination at a point in a room}}{\text{illumination under an unobstructed hemisphere of sky}} \times 100\%$$

So for instance, designers recommend a daylight factor of 2 per cent for a kitchen, 6 per cent for a drawing office, 2 per cent for school rooms, and so on. If we use for the illumination outside, a figure of 5000 lux, which has been found to be the average outside illumination in the United Kingdom over a number of years, and is called the *standard* or *design* sky, this gives a recommended figure of 100 lux for a kitchen and school room, and 300 lux for a drawing office.

When designers are doing daylight factor calculations, they measure the illumination in a room on the working plane, which is a plane or level (horizontal) at a height of 0.85 m above floor level.

Measurement of Daylight Factor

It is a fairly simple matter to measure the daylight factor in a room. Simply take a lightmeter, measure the illumination outside, then inside the room on the working plane, and substitute the two figures into the equation above.

A complication arises if we wish to know the illumination inside a room that has not yet been built. This is important because, having built it, we may discover that the room is not bright enough. The Building Research Station has developed a *Daylight Factor Protractor* and a *Daylight Factor Sliderule* especially for this purpose. Given these instruments, room and window sizes, and other relevant details, daylight factors for rooms and halls can be worked out before they have been constructed.

If a lightmeter is taken into a room and illumination levels measured at various positions, a contour map (figure 11.16) can be drawn up showing daylight factor variations around the room. These contours are useful when designing windows.

Artificial Lighting

Unlike daylight, which depends so much on the weather, artificial

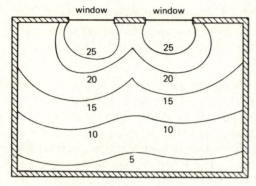

Figure 11.16

A great deal of technology is now available on light fittings. These are basically of three types, illustrated in figure 11.17.

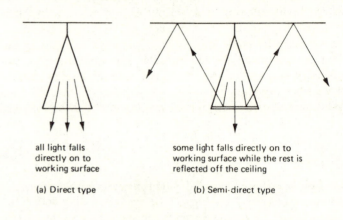

(a) Direct type — all light falls directly on to working surface

(b) Semi-direct type — some light falls directly on to working surface while the rest is reflected off the ceiling

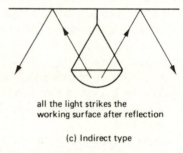

all the light strikes the working surface after reflection

(c) Indirect type

Figure 11.17

lighting is much more controllable. Recommended illumination levels using artifical light are given in table 11.1.

Table 11.1

Situation	Illumination (lux)
Casual use	100
General office work	400
Drawing office work	600
Fine detail, e.g. sewing	800
Minute detail, e.g. watchmaking	2000

The purpose of artificial lighting is to make the illumination on the working place as uniform as possible by spacing lights symmetrically. Under the *British Zonal System*, which is a method used for classifying light fittings, spacing/height ratios, and other information is available.

Too little illumination is undesirable, as is too much, which produces glare, leading to eyestrain, considered to be due to the eye looking and trying to adjust to brightly and dimly illuminated surfaces simultaneously.

Wall and Ceiling Surfaces

Even though a room or hall may have adequate windows and good artificial lighting, the surfaces of the walls and ceilings are also important. Light colours have higher reflection factors than dark colours, that is, light colours reflect more light. A room with black walls and ceiling is very unsatisfactory. A light, well-lit room contains a predominance of light pastel colours. Walls directly opposite windows should preferably be painted a light colour to give maximum light reflection. Gloss paints generally have higher reflection factors than flat, eggshell and satin finishes.

Some Unusual Properties of Light

Some of the more usual properties of light such as reflection and refraction are well known, and will not be mentioned here. The less well-known phenomena include *interference*, *diffraction* and *polarisation*.

Interference

Interference was first demonstrated by Young in 1801 and proved the wave nature of light. If two rays of light from the same coherent source are in phase, then the light intensity is a maximum. If they are out of phase then the intensity is less and may be a minimum if they are completely out of phase (see figure 11.18).

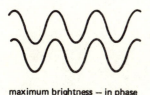
maximum brightness — in phase

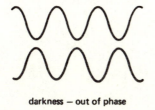
darkness — out of phase

Figure 11.18

Diffraction

Light is rectilinear, that is, it travels in straight lines, but on passing round an object or obstacle, it appears to bend or deviate. This phenomenon is called diffraction, and may be explained by wave theory, therefore lending support to the wave theory of light.

Polarisation

Light consists of transverse wave vibrations in all planes. If the vibrations occur only in one plane, the light is said to be *plane polarised*. Some substances will actually rotate the plane of polarised light. The use of polarisation in construction is that some materials, particularly polythene, will rotate the plane of polarisation when under stress. Thus, when a new structure is being planned, models of it may be made from polythene, which when under simulated stress, and subject to the scrutiny of polarised light, will show up the stress positions. The model, and the subsequent structure, can then be strengthened and modified in the appropriate positions.

EXERCISES

11.1 Define the following
   (a) the lux
   (b) the lumen
   (c) the candela
   (d) luminous intensity
   (e) flux.

11.2 What is meant by
   (a) colour rendering of a light source
   (b) incandescent light source?

11.3 What is meant by
   (a) daylight factor
   (b) daylight contours
   (c) working plane
   (d) daylight factor sliderule?

11.4 A lamp of luminous intensity 250 candelas is suspended 2.5 m above a round table of area 12.0 m^2. Calculate the illumination
   (a) in the centre of the table
   (b) at the edge of the table.

11.5   A lamp of 900 candelas is suspended 6 m above a road. What is the illumination of the road surface
  (a) at a point A directly under the lamp
  (b) at a point B on the road 8 m from A?

11.6   A lamp of 200 candelas is mounted 1 m below a large horizontal mirror. Determine the illumination at a point on a table 4 m below the lamp and 3 m to one side of the vertical through the lamp, if the mirror reflects 75 per cent of the light falling on to it.

11.7   Give *eight* factors you would take into account when providing a lighting scheme for any room or building using artifical lighting.

11.8   A detached single-room building is to be constructed as a drawing office. The interior is to have a high degree of illumination. Write down any *five* factors which a designer would have to bear in mind to achieve this.

11.9   What is polarised light? State *two* applications of it in building or structural engineering.

12 Heat Losses from Buildings

It is a well-known fact that heat always flows from a hot to a cold body. A house is heated to make it more comfortable, in other words, to raise its temperature above that of the outside environment. This means that because the house is at a higher temperature than its surroundings, it loses heat to the surroundings (figure 12.1). Unless the heat is replenished, the house will get cold, so more fuel is necessary to maintain a reasonable temperature inside. Fuel bills can be reduced somewhat by the introduction of insulating materials such as plastic foams and fibreglass, which delay, but cannot entirely prevent the house losing heat.

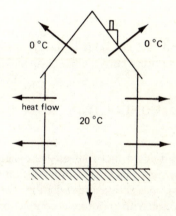

Figure 12..1

NEWTON'S LAW OF COOLING

On a cold day you find that you have to make up the fire or boiler more frequently than on a mild day. This is because heat is being lost at a faster rate. It was discovered by Newton that the rate of heat loss depends on the difference in temperature between the hot body and its surroundings. Thus house A in figure 12.2 loses heat more quickly than house B because of the greater temperature difference.

Expressed more formally, Newton's law of cooling states that

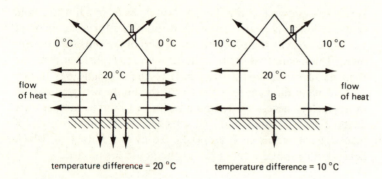

temperature difference = 20 °C   temperature difference = 10 °C

Figure 12.2

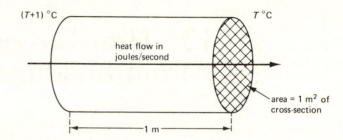

Figure 12.3

the rate of cooling of a hot body is directly proportional to the difference in temperature between the body and its surroundings.

CONVECTION, CONDUCTION AND RADIATION IN BUILDINGS

It is another well-known fact that all heat energy travels, or is transmitted, from one place to another by three different processes: convection, conduction and radiation. The most important of these in connection with buildings is conduction. A great deal of heat is lost from buildings by conduction through the fabric of the building: the walls, ceilings, roofs, floors, windows and doors. This constitutes the largest proportion of the heat loss.

Convection occurs only in liquids and gases—in this case air. Warm or hot air rises in rooms and leaves the building through windows and other vents. This is replaced by cold air which has to be warmed up. This cycle, although it wastes heat, is necessary to keep the air fresh. The amount of heat lost by this means is substantial enough to warrant calculation (see p. 87).

As regards radiation, which is transmission of heat via waves of electromagnetic radiation like light, the amount radiated depends on the temperature of the hot body. In this case the 'hot' body is the building, which being relatively cool, emits so little heat by radiation that it can be neglected.

Coefficient of Thermal Conductivity, k

Since heat is being lost from buildings, it is obviously important to know just how much. To put the subject on a quantitative basis, a coefficient of thermal conductivity has been introduced, which allows measurement of heat losses by conduction.

The coefficient of thermal conductivity of a substance, k, is the number of joules of heat flowing each second across the opposite faces of a material 1 m long and of cross-sectional area 1 m^2, the faces being maintained at a temperature difference of 1 °C (see figure 12.3). This can be expressed mathematically as

$$k = \frac{\text{rate of flow of heat}}{\text{cross-sectional area} \times \text{temperature gradient}}$$

where rate of flow of heat is measured in joules per second, cross-sectional area in m^2, and temperature gradient (which is the difference in temperature per unit length) is measured in °C per metre. The units of k are

$$\frac{\text{joule/second}}{\text{metre} \times \text{metre} \times °C/\text{metre}}$$

that is

$$\text{joule/second/metre/}°C$$

or

watt/metre °C

A few *k* values are given in table 12.1.

Table 12.1

Material	*k* value (W/m °C)
Copper	380
Iron	84
Brick	1.15
Plaster	0.50
Timber	0.12
Expanded polystyrene	0.03

Notice from the table that good conductors such as copper have high *k* values, whereas poor conductors have low *k* values.

Many practical methods are available for measuring *k*. Most of the materials used in house construction have low *k* values.

Thermal Transmittance or U Values

By means of *k* values, it is a fairly simple matter to calculate the flow or loss of heat through a material. However, in a building, the calculation of heat losses is complicated by the fact that most of the fabrics involved are not single materials. For instance, walls are made of brick, but they are usually plastered on one side, and may also be cement rendered on the other side. As the heat travels through the wall, it moves at different rates through the cement, brick and plaster, because they have different *k* values. This makes heat-loss calculations difficult.

In order to overcome this problem, another value, the *U* or thermal transmittance value has been introduced for composite materials, such as the wall mentioned above. So *k* values apply to single materials, and *U* values to composite structures.

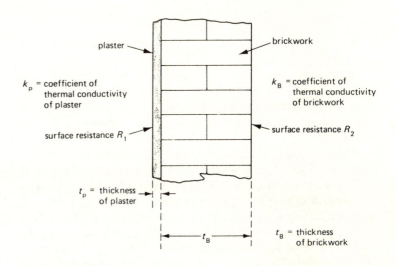

Figure 12.4

Calculation of U Values

U values may be calculated from *k* values in the following manner. Consider the composite wall structure shown in figure 12.4. We need to know the thermal transmittance, or *U* value, of the above wall, knowing the *k* values of plaster and brick.

Step 1 On reciprocating the *k* values, we obtain what are known as the *resistivity values*, thus

$$\text{resistivity of plaster} = \frac{1}{k_P}$$

$$\text{resistivity of brickwork} = \frac{1}{k_B}$$

Step 2 The resistivity values are converted to *resistance values* by multiplying by the respective thickness (in metres) of the material. Thus

$$\text{resistance of plaster} = \frac{1}{k_P} \times t_P$$

$$\text{resistance of brickwork} = \frac{1}{k_B} \times t_B$$

Step 3   The resistances worked out in step 2 give some measure of the resistance offered to heat flow by the respective materials concerned.

Another type of resistance to heat flow must be introduced at this stage. This is the *surface resistance*, which occurs on each side of the wall or structure; this is because the surface has some effect on the flow of heat through a structure. For instance, if the wall exterior surface were painted black, it would tend to absorb heat from the sun, and its temperature would rise, effectively reducing the temperature gradient and consequently the rate of heat flow. If, on the other hand, the wall faced north, and was therefore subject to cold winds, it would have a reduced temperature which would encourage heat flow through it.

At this stage all the resistances are added together, giving a *total resistance*, thus

$$\text{total resistance} = \begin{matrix}\text{interior}\\\text{surface}\\\text{resistance}\end{matrix} + \begin{matrix}\text{resistance}\\\text{of plaster}\end{matrix} + \begin{matrix}\text{resistance}\\\text{of brick}\end{matrix} + \begin{matrix}\text{exterior}\\\text{surface}\\\text{resistance}\end{matrix}$$

$$= R_1 + \frac{1}{k_P} t_P + \frac{1}{k_B} t_B + R_2$$

Step 4   When the total resistance is reciprocated, the result is called the *thermal transmittance* or *U* value, thus

$$U \text{ value} = \frac{1}{\text{total resistance}}$$

Example 12.1

Calculate the *U* value of the wall shown in figure 12.5.

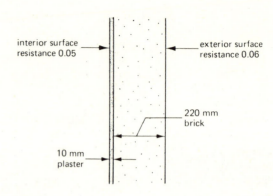

Figure 12.5

Solution

$$k_B = 0.8$$

$$k_P = 0.5$$

$$\begin{matrix}\text{Total}\\\text{resistance}\end{matrix} = \begin{matrix}\text{interior}\\\text{surface}\\\text{resistance}\end{matrix} + \begin{matrix}\text{resistance}\\\text{of}\\\text{brickwork}\end{matrix} + \begin{matrix}\text{resistance}\\\text{of}\\\text{plaster}\end{matrix} + \begin{matrix}\text{exterior}\\\text{surface}\\\text{resistance}\end{matrix}$$

$$= 0.05 + \frac{1}{0.8} \times 0.22 + \frac{1}{0.5} \times 0.01 + 0.06$$

$$= 0.05 + 0.28 + 0.02 + 0.06$$

$$= 0.41$$

$$U \text{ value} = \frac{1}{\text{total resistance}}$$

$$= \frac{1}{0.41} = 2.44$$

Therefore the *U* value of the wall = 2.44 W/m^2 °C.

Surface Resistance Values

It was stated above that the flow or transmission of heat through a structure, particularly walls and roofs, depends partly on the nature of the surfaces. For instance, a north-facing wall will be more heat conducting than a south-facing wall because it is colder, and so the temperature gradient is greater.

The figures in tables 12.2 and 12.3 are available for surface resistances.

Table 12.2 Internal surface resistances

Surface	Value (m^2/°C/W)
Walls	0.123
Roofs	0.104
Floors/ceilings	
heat flowing down	0.148
heat flowing up	0.104

Table 12.3 External surface resistances

Surface	Sheltered	Normal Exposure	Severe Exposure
South wall	0.128	0.100	0.078
West wall			
South-west wall	0.100	0.077	0.053
South-east wall			
North wall			
North-east wall	0.078	0.053	0.012
East wall			
North-west wall	0.078	0.053	0.031
All roofs	0.070	0.043	0.017

Table 12.4

Value	Unit
Conductivity = k value	watts/m °C
Resistivity = $\dfrac{1}{k}$	m °C/W
Resistance = $\dfrac{1}{k}$ × thickness	m^2 °C/W
U value = transmittance $= \dfrac{1}{\text{resistance}}$	W/m^2 °C

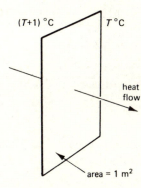

Figure 12.6

The Units of U Value

Notice, then, that from the units, the U value is effectively the number of watts, or joules per second, flowing across an area of 1 square metre of material, maintained at a 1 °C temperature difference (figure 12.6).

Example on Double Glazing

The U value of any composite structure can be calculated in the same manner as example 12.1, whether it is a wall, floor, ceiling or anything else. This example shows what saving can be obtained by introducing double glazing into a building. Let us assume the data given in figure 12.7 for a single-glazed window.

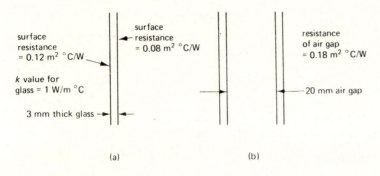

Figure 12.7

Solution   Single-glazed window

$$\text{total resistance} = \frac{\text{resistance}}{\text{of surface}} + \frac{\text{resistance}}{\text{of glass}} + \frac{\text{resistance}}{\text{of surface}}$$

$$= 0.12 + \frac{1}{1} \times 0.003 + 0.08$$

$$= 0.203$$

$$U \text{ value} = \frac{1}{\text{total resistance}} = \frac{1}{0.203} = 4.92 \text{ W/m}^2 \text{ °C}$$

Double-glazed window

$$\frac{\text{total}}{\text{resistance}} = \frac{\text{resistance}}{\text{of surface}} + \frac{\text{resistance}}{\text{of glass}} + \frac{\text{resistance}}{\text{of air gap}} + \frac{\text{resistance}}{\text{of glass}}$$

$$+ \frac{\text{resistance}}{\text{of surface}}$$

$$= 0.12 + \frac{1}{1} \times 0.003 + 0.18 + \frac{1}{1} \times 0.003$$

$$+ 0.08$$

$$= 0.12 + 0.003 + 0.18 + 0.003 + 0.08$$

$$= 0.386$$

$$U \text{ value} = \frac{1}{\text{total resistance}} = \frac{1}{0.386} = 2.59 \text{ W/m}^2 \text{ °C}$$

This means that in terms of a 1 square metre surface maintained at 1 °C temperature difference, 4.92 joules are lost each second through the glass. On introducing double glazing, 2.59 joules per second are lost. This represents a saving of 4.92 − 2.59 = 2.33 joules per second, that is, a (2.33/4.92) × 100 per cent or 47.3 per cent saving in heat loss.

With calculations of this sort, the value of double glazing can be readily appreciated.

CONVECTION HEAT LOSSES IN BUILDINGS

As we said earlier, all heat is lost from buildings either by conduction or convection. What has been said about U values applies to conduction losses; convection losses are calculated in a somewhat different way.

The air in a room becomes heated, but since the atmosphere must remain fresh, ventilation is provided, and as a result the warm air escapes. Cold air coming in to replace the warm air needs heat, and this is provided by the heating system present in the room or building. The problem of calculation resolves itself into one of finding how much heat is required for so many roomsfull of air lost in a given time (usually per second because the watt or joule per second is a very convenient unit to use).

It can be shown fairly simply that

$$\begin{array}{l}\text{heat lost} \\ \text{by} \\ \text{convection}\end{array} = \begin{array}{l}\text{volumetric} \\ \text{specific heat} \\ \text{of air} \\ \text{(constant at} \\ \text{1341 J/m}^3 \text{ °C)}\end{array} \times \begin{array}{l}\text{volume of} \\ \text{air in} \\ \text{room}\end{array} \times \begin{array}{l}\text{temp. difference} \\ \text{between incoming} \\ \text{and outgoing} \\ \text{air}\end{array}$$

$$\times \frac{\begin{array}{c}\text{number of air} \\ \text{changes per hour}\end{array}}{3600} \tag{12.1}$$

This can be expressed more concisely as

$$\text{heat lost (W)} = 1341 \times \text{volume of room} \times \text{temperature difference}$$

$$\times \frac{\text{number of air changes per hour}}{3600}$$

Example 12.2

A room has dimensions 5 m × 3 m × 2.5 m, and is to be maintained at a temperature of 20 °C. External fresh air is at 10 °C. If two air changes per hour are necessary for ventilation, calculate the heat loss.

Solution

$$\text{Heat loss} = 1341 \times (5 \times 3 \times 2.5) \times (20 - 10) \times \frac{2}{3600}$$

$$= 279 \text{ J/s}$$

TOTAL HEAT LOSS FROM A BUILDING

The heat lost by conduction through the fabric of a building requires a knowledge of the U value, which is the rate of heat loss through a square metre area of fabric with faces maintained at 1 °C temperature difference. Equation 12.2 expresses the heat lost through any area of fabric with faces maintained at any temperature difference

$$\begin{matrix} \text{heat lost by} \\ \text{conduction} \\ \text{through a} \\ \text{fabric} \end{matrix} = \begin{matrix} U \text{ value} \\ \text{of} \\ \text{fabric} \end{matrix} \times \begin{matrix} \text{area} \\ \text{of} \\ \text{fabric} \end{matrix} \times \begin{matrix} \text{temperature difference} \\ \text{between the faces of} \\ \text{the fabric} \end{matrix} \quad (12.2)$$

The heat lost by conduction (equation 12.2) is sometimes called *fabric heat loss*, and the heat lost by convection (equation 12.1) is called the *infiltration heat loss*, therefore

$$\begin{matrix} \text{total heat} \\ \text{loss} \end{matrix} = \begin{matrix} \text{fabric heat} \\ \text{loss} \end{matrix} + \begin{matrix} \text{infiltration} \\ \text{heat loss} \end{matrix}$$

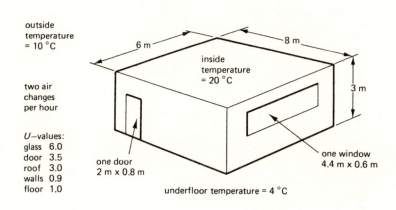

Figure 12.8

Example 12.3

Calculate the total heat loss from the building shown in figure 12.8.
Solution: Fabric heat loss (F.H.L.): The heat loss in each fabric of the building is calculated from the formula

$$\text{fabric heat loss} = U \text{ value} \times \text{temp. difference} \times \text{area of fabric.}$$

This is most simply presented in tabular form

	U value	Temp. Difference	Area	F.H.L. (W)
Door	3.5	10	1.6	56
Window	6.0	10	2.64	158.2
Roof	3.0	10	48	1440
Floor	1.0	16	48	768
Walls	0.9	10	79.8	718
			Total	3140.2

Infiltration heat loss (I.H.L.): The heat loss by ventilation is calculated from the formula

$$\text{Infiltration heat loss} = 1341 \times \frac{\text{volume of}}{\text{room}} \times \frac{\text{temp.}}{\text{difference}}$$

$$\times \frac{\text{no. of air changes/h}}{3600}$$

$$= 1341 \times (8 \times 6 \times 3) \times 10 \times \frac{2}{3600}$$

$$\text{I.H.L.} = 1072$$

$$\text{Total heat loss from building} = \text{F.H.L.} + \text{I.H.L.}$$
$$= 3140.2 + 1072$$
$$= 4212.2 \text{ W(J/s)}$$

Therefore, at least four 1 kW electric radiators would be needed.

The above example is obviously rather simple and theoretical, but the same technique can be and is used on calculations involving more complex real-life structures.

The total heat loss from our simple building is 4212.2 joules per second. If this quantity of heat is not replenished at the same rate, the interior temperature will fall; if the amount of heat replenished is greater, the temperature inside the building will rise.

It is a simple matter to make calculations for heating systems, such as radiator sizes, knowing the heat output required. Suppose, for example, that the room is to be heated by a radiator having an output of 4.2 watts per square metre area of radiator per degree centigrade temperature difference between radiator and surroundings, that, 4.2. $\text{W/m}^2 \text{ }^\circ\text{C}$ temperature difference. (The output of a radiator is often given in relation to the difference in temperature between radiator and room, because it obeys Newton's law of cooling, and therefore the greater the temperature difference, the greater the heat output.)

So, for a radiator temperature of 70 °C, that is, a difference of $70 - 20 = 50$ °C relative to room temperature, the heat output per

square metre of radiator $= 4.2 \times 50 = 210$ W. Therefore the radiator area needed to replenish the total loss of 4212.2 W $= 4212.2/210 = 20.0 \text{ m}^2$.

If we also know the *calorific values* of fuels, we can work out amounts of fuel required and at the same time the cost of such fuels.

EXERCISES

12.1 Composite structures have U values, but single materials such as plastics have k values. Explain why this is the case. Show how the U value for a composite structure may be obtained, given the necessary k values and surface resistances.

12.2 What is the importance of k values in building construction? Indicate how the values may be used in calculating heat losses and other factors.

12.3 What is meant by
(a) k value, coefficient of thermal conductivity
(b) U value, coefficient of thermal transmittance?

12.4 Why is a knowledge of U values important in building, and how are these values obtained?

12.5 (a) Calculate the thickness of insulative material that will reduce the heat loss by 80 per cent when fixed against the inside surface of the solid well described below.

	k value (W/m °C)
0.1 m brick	1.2
0.12 m stone	1.8
0.12 m plaster	0.58
R_{s_1}	0.123 m²/ °C/W
R_{s_2}	0.053

k for insulating material is 0.034 W/m °C

(b) If the average heat loss from a building is 66.8 kJ/s for the whole of the heating season, calculate the total cost of fuel used in replacing this heat loss for the heating season (30 weeks continuously). Efficiency of heating system is 60 per cent, the calorific value of fuel is 25.6 MJ/kg, and its cost is £20 per tonne.

12.6 (a) Determine the U value for a brick cavity wall comprising 220 mm brickwork, 12 mm plaster, 80 mm cavity. k values are brickwork 1.2 W/m °C, plaster 0.6, cavity resistance 0.18 m°C/W. $R_{s1} = 0.12$. $R_{s2} = 0.06$.
(b) Calculate the percentage saving in heat loss if 25 mm of polystyrene foam ($k = 0.04$ W/m °C) is fixed to the wall.
(c) If the reduction in heat loss from a building due to thermal insulation is 880 000 MJ per heating season, calculate the heating cost saved. Calorific value of fuel is 42 MJ/kg and the cost £0.0025 per kg, efficiency of heating system is 60 per cent.

12.7 A cavity wall is constructed of 0.1 m thick aerated concrete blocks used internally and 0.1 m thick solid concrete blocks used externally. The internal aerated blocks are covered with plaster to a thickness of 0.015 m and the external solid concrete blocks are covered with rendering 0.02 m thick. The thermal properties of the cavity wall are as follows.

Material	k value (W/m °C)
Solid concrete blocks	1.20
Aerated concrete blocks	0.15
Plaster	0.70
Rendering	1.00
	Resistance
Internal surface layer R_{s1}	0.12 m² °C/W
External surface layer R_{s2}	0.08
Airspace R_a	0.18

From this information determine the U value of the structure.

12.8 Calculate the total heat loss from the simple building shown in figure 12.9 using the information provided. Calculate also the surface area of radiators required to replenish the heat lost and maintain the room at a constant temperature. U values: window 5.00, walls 1.25, door 0.75, roof 3.50, floor 2.75. Outside air temperature = 5 °C; temperature under floor = 7 °C; inside air temperature = 20 °C; number of air changes per hour = 2; volumetric specific heat = 1341; heat output from radiators = 135 W/m².

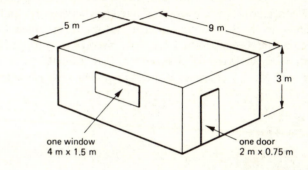

one window
4 m x 1.5 m

one door
2 m x 0.75 m

Figure 12.9

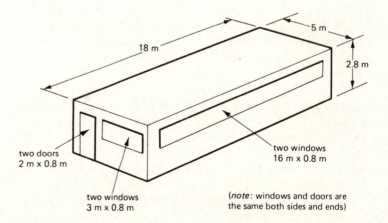

two doors
2 m x 0.8 m

two windows
16 m x 0.8 m

two windows
3 m x 0.8 m

(*note*: windows and doors are the same both sides and ends)

Figure 12.10

12.9 Determine the surface area of hot-water radiators and weight of fuel required to maintain an air temperature of 20 °C in the simple building in figure 12.10, through a heating season of 4000 hours.

Average temperature of air outside = 6 °C; average temperature under floor = 12 °C; number of air changes per hour = 2. Amount of heat required to raise the temperature of 1 m³ air by 1 °C = 1341 J. U values: windows 6.0, doors 2.8, walls 2.2, floor 2.4, roof 3.1. Heat output from radiators = 9 W/m² °C temperature difference (average temperature difference between the radiators and air in the building); heat output from pipework = 5.6 W/m run °C temperature difference; average temperature of radiators = 70 °C; length of pipework = 22 m; calorific value of the fuel = 36 MJ/kg; efficiency of heating system = 50 per cent.

13 Water Vapour and Humidity

The purpose of building science is to investigate ways of establishing and maintaining comfort in buildings—every chapter in this book corroborates this fact; this chapter on water vapour is similarly concerned with environmental comfort.

Although water vapour cannot be seen, there is usually plenty of it in the atmosphere, due to evaporation of water from seas, lakes and so on.

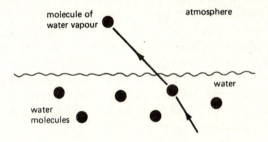

Figure 13.1

Molecules of water are, according to the kinetic theory, in constant random motion, and sometimes, in collisions with each other, they are knocked completely out of their liquid environment into the surrounding atmosphere, as shown in figure 13.1. The greater the water temperature, the greater is the rate of evaporation because the motion of the molecules increases and more are projected into the atmosphere.

All molecules of gas or vapour exert pressures on surfaces, and water vapour is no exception. The pressure of water vapour is known as its vapour pressure, and its magnitude is temperature dependent.

HUMIDITY AND HUMAN COMFORT

The human body has a marvellous central-heating system. Food is its fuel which, on being treated, liberates heat. The control of this heat is so good that the body temperature is always approximately 98.4 °F in a healthy body. Obviously there must be some sort of equilibrium process in operation because if heat is being continuously generated, and yet the body temperature remains constant, heat must be leaving the body at the same rate as it is being produced. Much of this heat is carried away in water vapour. Tiny drops of water emerge from the pores on to the surface of the skin, absorb heat, and vaporise away from the body (see figure 13.2). However, if the surrounding atmosphere is very humid, that is, there is already much water vapour present, this vaporisation process from the body is restricted or even stopped completely, because there is no more space available in the atmosphere for water vapour. (The atmosphere is saturated with water vapour.) The result of this is that the body tends to become overheated, and it feels 'stuffy'. In buildings, this problem is overcome either simply by opening a window, or by introducing air conditioning.

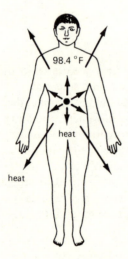

heat dissipated at same rate
as generated to get constant
temperature of 98.4 °F

Figure 13.2

HUMIDITY AND CONDENSATION

Just as water can evaporate to give water vapour in the atmosphere, so water vapour can condense back again to water. Condensation occurs when the atmosphere is saturated with water vapour. This is a problem when it occurs in buildings, and can cause serious trouble. It takes place when the air temperature drops. In the cooling process the air contracts and eventually becomes saturated. The temperature at which the atmosphere becomes saturated and condensation takes place is called the *dew point*.

Relative Humidity

It will readily be appreciated that the atmosphere is dry when there is not much water vapour present, and humid on a damp wettish day. The term *relative humidity* has been introduced to put the subject on a quantitative basis, where

$$\text{percentage relative humidity} = \frac{\text{amount of water vapour in the air} \times 100}{\text{amount of water vapour necessary to saturate the air}}$$

Hygrometry

A *hygrometer* is an instrument used to measure relative humidity. There are many different types available, some of which will be dealt with below: wet and dry bulb hygrometer, hair hygrometer, paper hygrometer and Regnault's dewpoint hygrometer.

Wet and Dry Bulb Hygrometer

This is the most accurate hygrometer (see figure 13.3). It consists of two thermometers, one with a dry bulb, and the other with a damp or wet bulb, covered with a water-soaked cloth, such as muslin. The wet bulb thermometer gives a lower reading than the dry bulb, because as water evaporates from the muslin, it takes heat with it, keeping the wet bulb cool. The difference between the readings, *H*,

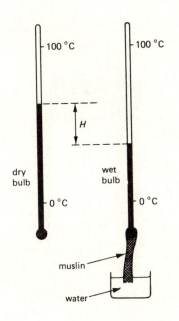

Figure 13.3

is related to the relative humidity, which may be found by reference to table 13.1.

The amount of cooling of the wet bulb depends on the rate of water evaporation from the muslin, and this in turn depends on the humidity of the surrounding air. If the surrounding air is very humid, then evaporation from the muslin is very limited because the air cannot take much more water vapour. The cooling of the wet bulb is therefore only slight, and the two thermometers read almost the same. If the air is completely saturated, no evaporation takes place from the wet bulb, and so $H = 0$.

If on the other hand, the humidity of the atmosphere is very low, that is, the air is dry, a great deal of evaporation occurs around the wet bulb, and H is large.

Sahara Desert: 0% relative humidity, H large (maximum)
Sauna bath   : 100% relative humidity, $H = 0$ (minimum)

Note To obtain results quickly, a whirling or rotating type is also available.

Table 13.1 Tables for Wet and Dry Bulb Hygrometer

Dry Bulb Thermometer Reading	Difference between Readings (°F)																	
	1	2	3	4	5	6	7	8	9	10	11	12	13	14	15	16	17	18
	Relative Humidity (%)																	
35	90	81	72	64	57	51	45	39	35	32	28	25	22	19	17	15	13	11
40	92	84	76	70	68	58	52	47	43	38	34	31	28	25	23	21	19	17
45	92	84	78	72	65	60	55	50	46	41	38	34	31	28	25	23	21	19
50	93	86	79	73	68	62	58	53	49	45	41	38	34	32	29	26	24	21
55	93	86	80	75	69	64	59	55	51	47	44	40	37	34	31	29	26	24
60	93	88	82	76	71	66	62	58	54	50	46	43	40	37	35	32	30	27
65	94	88	83	78	73	68	63	59	55	52	48	45	42	39	36	34	31	29
70	94	89	83	78	74	69	65	61	58	54	50	47	44	41	38	36	34	31
75	94	89	84	79	74	70	66	63	59	55	52	49	46	43	40	38	35	33
80	95	90	85	80	76	72	67	64	60	57	53	50	47	45	42	39	37	35
90	95	90	85	81	77	73	69	65	62	59	56	53	50	47	44	42	40	38
100	95	90	86	82	78	74	70	67	64	61	58	55	52	49	47	45	43	41

Hair and Paper Hygrometers

Both of these hygrometers work on the principles that hair and paper increase in length when damp, and shorten when dry. Therefore, on a humid day the materials lengthen, whereas in dry weather they contract. The dimensional changes taking place in the hair and paper are caused to move pointers over the scales. Both types of instrument are direct reading, but neither is particularly accurate since both materials are somewhat inelastic. A typical paper hygrometer is shown in figure 13.4.

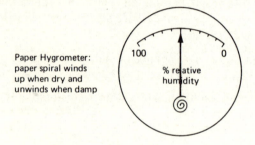

Paper Hygrometer: paper spiral winds up when dry and unwinds when damp

Figure 13.4

Regnault's Dewpoint Hygrometer (figure 13.5)

Air is blown through ether, which extracts heat from its surroundings as it evaporates. Eventually the air immediately around the bottom of the tube becomes saturated, and dew forms on the silvered surface. The temperature is noted at the point when dew forms. The apparatus is allowed to warm up again and the temperature noted when the dew disappears. The average of these two temperatures is the dewpoint. This equipment is not direct reading, and the humidity is calculated with reference to saturated vapour pressure tables and use of the following equation

$$\text{percentage relative humidity} = \frac{\text{saturated vapour pressure of water at dewpoint}}{\text{saturated vapour pressure of water at room temperature}} \times 100$$

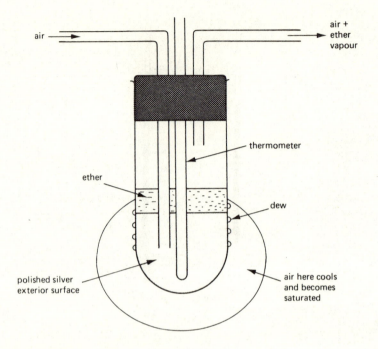

Figure 13.5

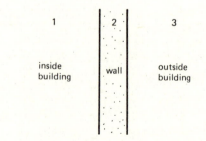

Figure 13.6

Condensation

Condensation can occur in three places as indicated in figure 13.6.

Condensation always occurs at the dewpoint, that is, the temperature at which the atmosphere becomes saturated with water vapour. The dewpoint temperature is not, of course, constant, but varies with the amount of water vapour present. The greater the amount of water vapour present, the higher the dewpoint temperature.

Outside the Building

In general, condensation outside presents few problems to a building because it is built to deal with rainwater anyway. During the night, the temperature drops, and under suitable conditions, dew forms, in other words, condensation occurs. This water falls to the ground, and re-evaporates again during the day when the temperature rises.

Inside the Building

Condensation can occur in two ways, and its presence is detected by the formation of water on cold surfaces such as windows and walls.

(1) It can occur during the night when the temperature in a room falls to the dew point, particularly that air closest to a window through which heat is easily lost.

(2) It can occur in a room where the humidity is increasing, for instance, in a bathroom where someone is having a bath, or in a small room crowded with people.

Condensation of water droplets on cold surfaces is undesirable for many reasons, as follows.

(1) Causes corrosion of metal pipes and window frames.
(2) Causes blistering and peeling of paintwork.
(3) Causes discolouration of wallpaper.
(4) Provides a suitable environment for mould and algae growth.
(5) Causes blistering in some plasters.
(6) Causes discomfort and unpleasantness generally.

To prevent condensation, we are presented with two alternatives. One is to keep the air dry, one of the simplest ways of doing this being to open a window, or else introduce air-conditioning. The second alternative is to maintain temperatures above the dewpoint, either by warming the room or providing local insulation to the surfaces concerned such as an expanded foam plastics layer applied to walls.

Within the Wall

Let us suppose that the interior temperature of a house is 20 °C, and the temperature outside 3 °C, as in figure 13.7. Suppose also that the dewpoint temperature happens to be 7 °C. This means that there is a temperature drop of 13 °C across the wall. Since bricks are porous and contain air, water vapour must be present within the wall, and at wall temperatures at and below 7 °C, condensation will take place. These condensed water droplets can move through the wall by capillary action and cause dampness. This potential defect may be remedied by introducing an impermeable membrane, called a *vapour barrier*, such as aluminium applied either as foil or paint to the warm surface of the wall. Aluminium-faced plasterboard is also available for this purpose.

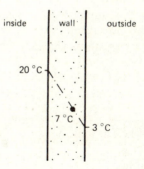

Figure 13.7

HOW TO BE COMFORTABLE IN A BUILDING

The reason why we build dwellings for ourselves, and arrange for them to be warm, well-lit, and so on, is to make life as comfortable as possible. Our comfort in a building depends on

(1) the temperature of the building
(2) adequate air and ventilation
(3) adequate lighting
(4) the humidity of the air in the building
(5) noise factors.

These are according to personal priorities—perhaps your priorities are different.

Temperature

Optimum levels have been established at 20 °C for sedentary workers and 12.5 to 15.5 °C for manual workers. In general, since the human skin is more sensitive to infrared or heat radiation than to other forms of heating, radiant heating is comfortable at a somewhat lower temperature (15.5 °C for sedentary workers instead of 20 °C).

Air and Ventilation

The normal oxygen content of the air is 20 per cent. If it drops to below 15 per cent the amount of oxygen is insufficient and causes discomfort. A slight movement of air is considered desirable, although obviously too much movement causes unpleasant draughts.

Lighting

This subject has been treated more thoroughly in chapter 11, but in general, the figures below indicate some levels of illumination required for comfort
restaurants 100 lux
study rooms 350 lux
drawing offices 500 lux

Humidity

Optimum levels are considered to be 50 to 60 per cent relative humidity. Below this the air is too dry, but above 60 per cent, it is too 'stuffy'.

Noise

The subject of noise levels has been treated in chapter 9. The threshold of hearing, which is the lowest sound level to be detected by the human ear is 1 decibel (dB). The threshold of pain is 120 dB. Above this value the eardrum can suffer permanent damage. A sound level of 60 dB and above is distracting.

EXERCISES

13.1 Explain what is meant by humidity and describe how it may be measured.

13.2 List and briefly explain the main factors affecting the environment in a building giving the limits of the conditions for comfort where applicable.

13.3 Give a concise account of the humidity of the air, and condensation control in buildings.

13.4 Give *four* methods you would use to reduce or prevent condensation.

13.5 Briefly describe *six* factors which affect the comfort of people in buildings.

13.6 An experiment to measure the saturated vapour pressure of water at various temperatures gave the following results.

Temperature (°C)	0	10	20	30	40	50	60
Vapour pressure (mm mercury)	4.6	9.2	17.5	31.8	55.3	92.5	149.4

Plot the results on a graph and use it to deduce the vapour pressure of water in a room where the dewpoint is 15 °C.

14 Applied Mechanics

STRESS, STRAIN AND ELASTICITY

A force applied to a material causes it to change its shape. The material may become longer, shorter, twisted or bent. This phenomenon has been investigated quantitatively, and gives rise to certain important terms. The *stress* experienced by the material, which depends on the size of the force operating, and the area over which it operates (figure 14.1) is given by the equation

$$\text{stress} = \frac{\text{force}}{\text{cross-sectional area}} \text{ N/mm}^2$$

Figure 14.1

If we take the case where, say, a bar is simply being lengthened or shortened, the change in length is a measure of the *strain* in the bar, where

$$\text{strain} = \frac{\text{change in length}}{\text{original length}} \text{(no units)}$$

Any material distorted by a force will return to its original shape when the force is removed, within certain limits. This behaviour is termed *elasticity*. All materials are elastic, but rubber obviously has much wider limits than, for example, a material such as concrete or steel. It was discovered that the change in length of a material is directly proportional to the force applied to it, within the elastic limit; this important relationship is called Hooke's law. When a material is distorted beyond its elastic limit, it will not return to its original shape when the distorting force is removed.

It was first recognised by Young that the stress in a material is proportional to its strain; and this gives rise to another important equation

$$\frac{\text{stress}}{\text{strain}} = \text{Young's modulus, } E$$

Nearly all materials subject to a force or load exhibit the behaviour shown in figure 14.2.

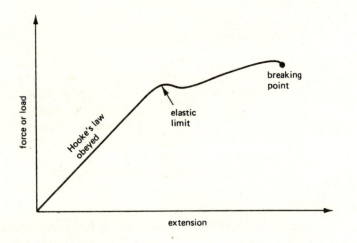

Figure 14.2

Materials in load-bearing structures are subject to stress and care has to be exercised so that this is not excessive. The normal working stress is related to the ultimate or breaking stress by means of a factor of safety, thus

$$\text{factor of safety} = \frac{\text{ultimate stress}}{\text{working stress}}$$

The purpose of this system is to allow a good margin of safety. For example, suppose the factor of safety for a load-bearing column is 3. This means that the normal allowable working stress within it will be only one-third of the value which would be required for the column to fail.

Example 14.1

A circular steel column 3 m long of 100 mm² cross-sectional area is subject to a load of 10 kN. Calculate the stress, strain and factor of safety if Young's modulus is 210 kN/mm² and the breaking stress is 350 N/mm².

Solution

$$\text{Stress} = \frac{\text{load}}{\text{cross-sectional area}} = \frac{10\ 000}{100} = 100 \text{ N/mm}^2$$

$$\text{Strain} = \frac{\text{stress}}{E} = \frac{100}{210\ 000} = 0.00048$$

$$\text{Factor of safety} = \frac{\text{breaking stress}}{\text{working stress}} = \frac{350}{100} = 3.5$$

PRINCIPLE OF MOMENTS

When a force is causing or tending to cause turning, it is said to have a turning moment. On the other hand, when a force is producing, or tending to produce, bending it is said to have a bending moment. Bending moments will be dealt with briefly later, but the principle of moments deals with turning moments. The size of the turning moment is the product of force and distance, measured normally from the force to the point of rotation. In figure 14.3

$$\begin{aligned} \text{turning moment} &= \text{force} \times \text{perpendicular distance} \\ &= 5 \times 3 \\ &= 15 \text{ N m} \end{aligned}$$

If the force is producing a clockwise tendency for rotation about a point, it is said to possess a clockwise moment. In the same way an anticlockwise motion produces an anticlockwise moment. The *principle of moments* states that for a system of forces in equilibrium the sum of the clockwise moments produced is equal to the sum of the anticlockwise moments about any fixed point.

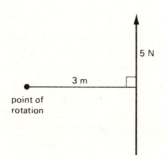

Figure 14.3

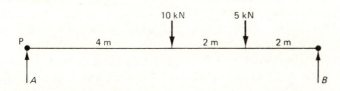

Figure 14.4

The principle is true for any system of forces but we shall restrict it to systems of parallel forces.

Application of the principle includes calculations of beam supports or reactions. For example, the reactions A and B may be worked out for the beam loaded as shown in figure 14.4. A point must initially be chosen to consider moments, and this is usually most convenient on the beam at one of the supports, say, point P. Then

sum of clockwise moments about P $= (10 \times 4) + (5 \times 6)$

and

sum of anticlockwise moments about P $= 8B$

Since clockwise moments = anticlockwise moments

$$(10 \times 4) + (5 \times 6) = 8B$$
$$B = 70/8 = 8.75 \text{ kN}$$

Also, since the total upward force must equal the total downward force for equilibrium

$$A = 10 + 5 - 8.75 = 6.25 \text{ kN}$$

Shear-force and Bending-moment Diagrams

Any person who is concerned seriously with construction studies must sooner or later acquire some knowledge of shear forces and bending moments. Here we shall just explain how the diagrams are drawn, and let it be sufficient to state that this knowledge subsequently becomes vital to anyone engaged in the structural design of buildings.

A beam in a building may fail in a number of ways including shearing and excessive bending. It is therefore important to know how the shear forces are distributed along a beam, and also its state of bending.

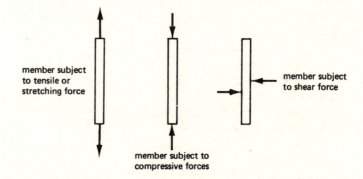

Figure 14.5

Three important modes of operation of forces on a structural member are illustrated in figure 14.5. Tensile forces obviously stretch members, compressive forces shorten them, while shear forces tend to produce failure or breakage of members at points between the two displaced equal forces.

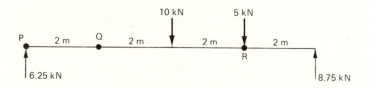

Figure 14.6

Let us consider the point-loaded beam shown in figure 14.6. If we take any point along the beam, the total force to the left of that point is equal to the total force to the right of the point, but the forces are in opposite directions. The size of force at this point is called the shear force. Thus at point Q, the total force is 6.25 kN up to the left, and $10 + 5 - 8.75 = 6.25$ kN down to the right. The shear force at Q is therefore 6.25 kN. If the shear force is calculated at all points along the beam in this way and the values plotted on a single diagram, the diagram is called a shear-force diagram, and it indicates the distribution of shear force along the beam. The shear-force diagram for the beam in figure 14.6 is shown in figure 14.7.

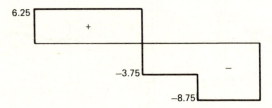

Figure 14.7

These diagrams may be drawn in other ways but the convention used here is that if the total force to the left of the point chosen is upwards, this is taken as positive, and negative if downwards.

The bending moment may also be calculated at any point along a beam, and it gives a measure of the state of bending of the beam at that point. The bending-moment diagram shows the distribution of this quantity along the beam.

Students often have difficulty in calculating bending moments, although they are simple if we remember the following procedure. Select a point on the beam and calculate the moment due to bending *either* by calculating the total moment to the left *or* the total moment to the right. The convention used here is that upward forces produce positive moments, and downward forces negative moments. For example, at point Q on the beam in figure 14.6 the total moment to the left of the point is $+ 6.25 \times 2 = + 12.5$ N m, while the total moment to the right of this point is $+ (8.75 \times 6) - (10 \times 2) - (5 \times 4) = 12.5$ N m. It can be seen that calculating either to right or left produces the same result. At point Q, the bending moment is $+ 12.5$ N m. To calculate the bending moment at P, again work either to left or right of this point, whichever is easier. Looking left, there are obviously no forces, and therefore the bending moment is zero. The same result would be achieved by calculating to the right of this point, thus $+ (8.75 \times 8) - (10 \times 4) - (5 \times 6) = 0$. The bending moment at R, calculated in the same way, is $+ 17.5$ N m. When enough points have been worked out, the bending-moment diagram can be drawn as shown in figure 14.8.

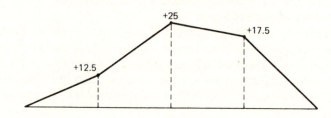

Figure 14.8

Another example is given in figure 14.9. Taking moments at P clockwise moments = anticlockwise moments

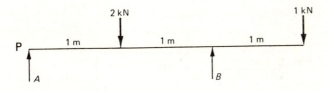

Figure 14.9

$$(2 \times 1) + (1 \times 3) = 2B$$
$$B = 2.5 \text{ kN}$$
It follows that $A = 0.5$ kN.

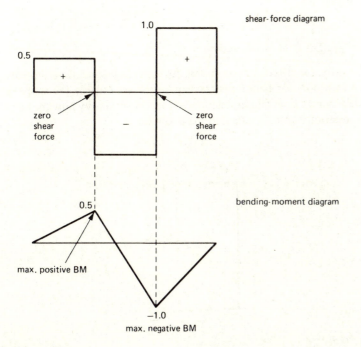

Figure 14.10

The purpose of the diagrams in figure 14.10 is to indicate important features, particularly points on the beam where the

state of bending is the greatest, that is, the maximum bending moment, and points where the shear force is least and greatest. Note that the points on the beam where the bending moment is a maximum correspond to points where the shear force is zero, and conversely, maximum shear force corresponds to zero bending. The former fact is very important and allows engineers to calculate the position of maximum bending moment without going to the trouble of drawing the bending-moment diagram, which, in complicated beams, can be very troublesome. All that they are required to do is to draw the shear-force diagram (which is simple) locate the position of zero shear force and calculate the bending moment at this point, where is has a maximum value. This gives vital information with regard to designing the beam.

Students are strongly advised to master this subject, and more examples are given for practice at the end of the chapter.

BEAM DESIGN

The design of simple columns may be carried out using the ideas and equations indicated earlier in this chapter under the section on stress, strain and elasticity. The design of beams is more complicated.

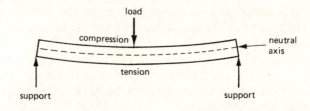

Figure 14.11

A beam obviously carries a load and this causes a certain amount of bending (figure 14.11). As a result of this, the lower part of the beam is in a state of stretch or tension, and the upper part in a state of compression. The centre of the beam is neither in tension nor compression and is called the *neutral axis* in this connection. From common sense it can be seen that the amount of tension or compression (and therefore the stress) increases with distance from the neutral axis. This variation is usually expressed on a stress-variation diagram (figure 14.12).

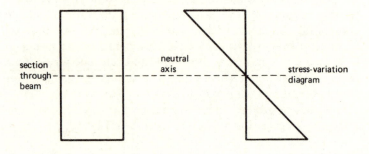

Figure 14.12

The maximum stress is therefore in the top or bottom fibres of the beam. This maximum value is important and is the criterion for designing the beam. There is, however, another very important factor to consider. The stress along a beam varies, and depends on the state of bending of the beam. The greater the bending, the greater is the stress. The greatest stress occurs at the position of maximum bending moment.

To sum up a little—it is vital to know the position and size of the maximum stress due to bending. The position of this maximum stress is at the top or bottom fibres of the beam, at the point of maximum bending moment.

Consider the following example illustrated in figure 14.13. A beam 4 m long is carrying a load of 10 kN at midspan. From the bending-moment diagram, the maximum bending moment is seen to be in the centre. The centre of the beam is therefore under the greatest stress, and the actual points of greatest stress, P and Q are in the extreme fibres. The numerical values of the stresses at P and Q are identical in simple beams, the only difference being that the stress at P is compressive, while that at Q is tensile.

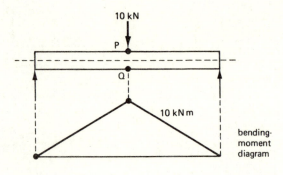

Figure 14.13

Calculation of Bending Stress

Various methods are available for calculating bending stresses, but proofs for bending stress equations are outside the scope of this book. It will be sufficient here to indicate simply how the basic equation for such calculations is obtained.

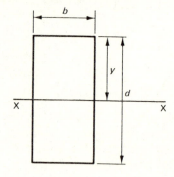

Figure 14.14

Now the bending stress is proportional to the bending moment, M and to the distance measured above or below the neutral axis, y (figure 14.14), that is

$$f \propto My$$
$$f = \text{constant} \times My$$

The constant here is equal to $1/I_{XX}$, where I_{XX} is called the *second moment of area*, and its value depends on the breadth and depth, d, of the beam. It can be shown that for a beam of rectangular section $I_{XX} = bd^3/12$. In physical terms I_{XX} may be regarded as the contribution the beam itself makes towards resisting the bending put into it by the applied load. It can be seen, for instance, that the greater the depth of the beam, the greater is its I_{XX} value, and the greater is its resistance to bending.

The above equation therefore becomes

$$f = \frac{1}{I_{XX}} \times My$$

or

$$\frac{M}{I_{XX}} = \frac{f}{y}$$

Normally, the maximum bending stress is required, and y is equal in this case to half the depth of the beam, and M is the maximum bending moment. This gives, on rearrangement

$$f_{max} = \frac{M_{max}}{I_{XX}} \times y$$

where f_{max} = maximum bending stress (N/mm²), M_{max} = maximum bending moment (N mm), I_{XX} = second moment of area (mm⁴) and y = distance from neutral axis to the top or bottom of the beam (mm).

To show how such calculations are done, consider the following examples.

Example 14.2

A beam of breadth 50 mm, depth 100 mm and length 5 m is simply supported at its ends, and carries a central point load of 2 kN

(figure 14.15). Calculate the maximum bending stress in the beam.

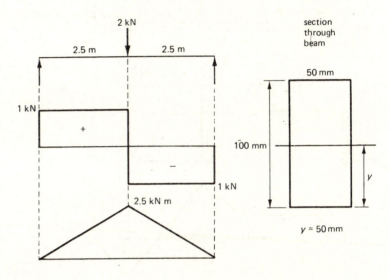

Figure 14.15

Solution

Second moment of area, $I_{XX} = \dfrac{bd^3}{12}$

$$= \frac{50 \times 100^3}{12}$$

$$= 4.166 \times 10^6 \text{ mm}^4$$

Maximum bending moment = 2.5 kN m
$$= 2.5 \times 10^6 \text{ N mm}$$

Maximum Bending Stress, $f_{max} = \dfrac{M_{max} \, Y}{I_{XX}}$

$$= \frac{2.5 \times 10^6 \times 50}{4.166 \times 10^6}$$

$$= 29.9 \text{ N/mm}^2$$

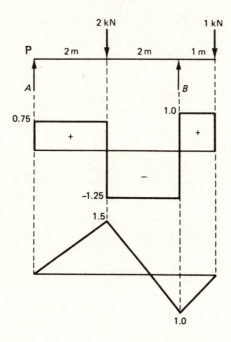

Figure 14.16

Example 14.3

The beam from example 14.2 is now loaded and supported as shown in figure 14.16. Calculate the maximum stress due to bending.

Solution To find *A* and *B* take moments about P.

Clockwise moments = anticlockwise moments
$(2 \times 2) + (1 \times 5) = 4B$
$B = 2.25 \, kN$
$A = 3 - 2.5 = 0.75 \, kN$
Maximum bending moment $= 1.5 \times 10^6 \, Nmm$

Maximum bending stress, $f_{max} = \dfrac{1.5 \times 10^6 \times 50}{4.166 \times 10^6}$

$\qquad\qquad = 17.9 \, N/mm^2$

EXERCISES

14.1 Define the terms
 (a) stress
 (b) strain
 (c) Young's modulus
 (d) Hooke's law
 (e) elastic limit
 (f) factor of safety
 (g) elasticity.

14.2 An axial load of 2500 N acts on a steel tie bar of length 5.5 m and area of cross-section 30 mm². Calculate the increase in length produced ($E = 210 \, kN/mm^2$).

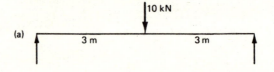

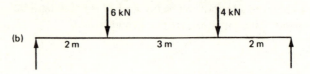

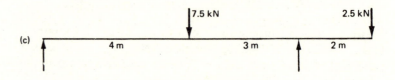

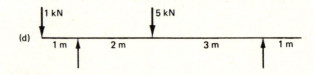

Figure 14.17

14.3 A piece of timber of square cross-section of side 25 mm fails under a load of 12 kN. Calculate the permissible stress if the factor of safety is 3.5.

14.4 For the point-loaded beams given in figure 14.17, calculate the reactions, draw the shear-force and bending-moment diagrams, and write down the position and size of the maximum bending moment in each case.

14.5 A beam of length 4 m, breadth 100 mm and depth 200 mm, is simply supported at its ends and carries a central point load of 10.5 kN. Calculate the maximum bending stress in the beam.

14.6 Calculate the sectional area of a square-section beam, if it is to be 5 m long, carry a central load of 15 kN, and the maximum stress within it must not exceed 20 N/mm^2.

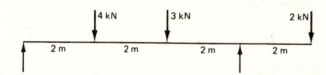

Figure 14.18

14.7 A beam of breadth 50 mm and depth 200 mm is supported and loaded as shown in figure 14.18. Calculate the maximum bending stress in it.

Answers to Numerical Questions

Chapter 3 Concrete (p. 26)

15.   (a) 52.5 kg,   (b) 237.5 kg,   (c) 0.379 m^3,
(d) 2451 kg/m^3, (e) 2360 kg/m^3

Chapter 4 Water (p. 34)

3. 5.06 m/s
4. 22.3 min

Chapter 9 Sound (p. 65)

2.   1.89 s, 762 people
3.   1.47 s
5.   22.0 m^2
6.   80.63 dB
7.   88.0 dB
8.   85.1 dB

Chapter 10 Electricity (p. 73)

1.   0.56 A
5.   0.40 A

Chapter 11 Light (p. 83)

4.   (a) 40.0 lux,   (b) 20.0 lux
5.   (a) 25.0 lux, (b) 5.4 lux
6.   9.38 lux

Chapter 12 Heat Losses from Buildings (p. 92)

5.   (a) 37.3 mm, (b) £1577.8
6.   (a) 1.79 W/m^2 °C, (b) 53%, (c) £87.2
7.   0.85 W/m^2 °C
8.   7378 W, 54.6 m^2
9.   17.5 m^2, 11.24 tonne

Chapter 13 Water Vapour and Humidity (p. 100)

6.   12 mm of mercury

Chapter 14 Applied Mechanics p. 108)

2.   2.145 mm
3.   5.48 N/mm^2
4.   (a) midpoint of beam, 15 kN m, (b) 2 m from left-hand end, 10.8 kN m, (c) 4 m from left-hand end, 10 kN m, (d) 3 m from left-hand end, 5.4 kN m
5.   15.75 N/mm^2
6.   31 600 mm^2
7.   18 N/mm^2